A TEXT BOOK ON PRODUCTION PROCESSES

Dr. Girisha L
Dr. R. Sanjeev kumar
Dr. Ram Subbiah
Dr. Raghu Chand R
Dr. Mahesh M

SHANLAX PUBLICATIONS

A Text Book on Production Processes

First Edition: December 2020

ISBN: 978-93-90082-91-9

Price: ₹345/-

Publisher

SHANLAX PUBLICATIONS
61, 66 T.P.K. Main Road
Vasantha Nagar
Madurai – 625003
Tamil Nadu, India

Ph: 0452-4208765,
Mobile: 7639303383
email:publisher@shanlaxpublications.com
web: www.shanlaxpublications.com

PREFACE

The objective of this book is to present the concepts and applications of production processes. The purpose of this book to make it easier to understand about Mechanical Engineering based production processes.

We have confident about this book which would be the very useful for UG students, Research Scholars and Post Graduate Engineering student of Mechanical engineering to understand the concepts and applications.

Though unrest care has been taken to check mistakes, yet it is difficult to claim perfection. The author will therefore, acknowledge any suggestions for the improvement of the book and will be incorporated is the next edition.

Dr. Girisha L
Dr. R. Sanjeev kumar
Dr. Ram Subbiah
Dr. Raghu Chand R
Dr. Mahesh M

CONTENTS

UNIT – I

CASTING PROCESS

1. INTRODUCTION TO CASTING PROCESS

Manufacturing Process In industries, many different processes or manufacturing methods are used. To be able to select the technically and economically best manufacturing sequence for a product, it is necessary to have a broad fundamental knowledge of the possibilities and limitations of the various manufacturing processes, including work materials used and geometries, surface finishes & tolerances required. The term process can in general be defined as a change in the properties of an object, including geometry, hardness, state, information content (form data), and so on.

Production or manufacturing can be simply defined as value addition processes by which raw materials of low utility and value due to its inadequate material properties and poor or irregular size, shape and finish are converted into high utility and valued products with definite dimensions, forms and finish imparting some functional ability. Manufacturing Process basically deals with the conversion of raw materials to end product or useful material (output or finished product or component).

However, manufacturing processes can be broadly classified in four major groups as follows:

(a) **Shaping or forming:** manufacturing a solid product of definite size and shape from a given material taken in three possible states: • solid state – e.g., forging, rolling, extrusion, drawing, etc. •liquid or semi-liquid state – e.g., casting, injection molding , etc. • powder form – e.g., powder metallurgical process.

(b) **Joining process:** Welding, brazing, soldering etc.

(c) **Removal process:** Machining (Traditional – turning, drilling, milling, shaping, grinding or Non-traditional – ECM, EDM, ECG, EBM, LBM, Ultrasonic Machining, photochemical Machining, etc.)

(d) **Regenerative manufacturing:** Production of solid products in layer by layer from raw materials in different form: (Rapid Prototyping & Manufacturing) • liquid – e.g., Stereo Lithography (STL) • powder – e.g., Selective Laser Sintering (SLS) • sheet – e.g., Laminated Object Manufacturing (LOM) • wire – e.g., Fused Deposition Modeling (FDM).

Casting processes basically involves the introduction of molten metal into a mold cavity, where upon solidification, the metal takes on the shape of the mold cavity. Simple and complicated shapes can be made from any metal that can be melted. Example of cast parts: frames, structural parts, machine components, engine blocks, valves, pipes, statues, etc.

Components are produced by pouring molten metal into a contoured (mould) cavity followed by cooling to a solid mass. Cooled solid mass represents the configuration of the cavity and is the required shape of the component. Components thus produced are referred to as castings. The cavity corresponding to the shape of the components is called as the mould. The mould may be made of refractory grains (sand moulds) or metal (metal moulds). The cooling of liquid metal to solid metal is termed phase transformation. The place where this activity is carried out is referred to as foundry.

2. BASIC TERMINOLOGY IN CASTING PROCESS

- **Mould (Mold or flask)** - is essentially a negative of the final product or having a cavity of the casting shape which is to be manufactured, in which molten metal is poured.
- **Pattern** - made of wood or metal, is a replica of the final product and is used for preparing mould cavity.
- **Core** – is an obstruction (made of either sand or metal) positioned in the mold does not permit molten metal to fill space occupied by core, thereby making holes or hollow cavities inside a casting.
- **MoldBox** – is a combination of two halves; upper half is called cope and lower half drag.
- **Moldmaterial** – is one out of which mold is made, must retain its shape till metal has solidified. Wax, plaster of Paris, ceramics, etc. are some materials.
- **Pouringcup** - is a funnel shaped cup which forms the top portion of sprue and allows operator to direct flow of metal from crucible to Sprue.
- **Sprue** – is a vertical channel (square or round) through which molten metal flows downward in the mold and enters the runner & gate.
- **Pouringbasin** – acts as a reservoir from which molten metal moves smoothly into sprue, holds back slag & dirt which floats on top.
- **Runner** – channels that carry the molten metal from the sprue to the mold cavity.
- **Gate** – is the portion of the runner through which the molten metal enters the mold cavity; more gates for large castings can be made.
- **Riser** – serve as reservoirs to supply the molten metal necessary to prevent shrinkage, permits escape of air & mold gases.
- **Ladle** – used to carry molten metal from the furnace to the molding boxes.

3. MOLDING SANDS

Molding sands, also known as foundry sands, are defined by eight characteristics: refractoriness, chemical inertness, permeability, surface finish, cohesiveness, flow ability, collapsibility, and availability/cost.

Types of base sands

Base sand is the type used to make the mold or core without any binder. Because it does not have a binder it will not bond together and is not usable in this state.

Silica sand

Silica (SiO_2) sand is the sand found on a beach and is also the most commonly used sand. It is made by either crushing sandstone or taken from natural occurring locations, such as beaches and river beds. The fusion point of pure silica is 1,760 °C (3,200 °F), however the sands used have a lower melting point due to impurities. For high melting point casting, such as steels, a minimum of 98% pure silica sand must be used; however for lower melting point metals, such as cast iron and non-ferrous metals, a lower purity sand can be used (between 94 and 98% pure).

Silica sand is the most commonly used sand because of its great abundance, and, thus, low cost (therein being its greatest advantage). Its disadvantages are high thermal expansion, which can cause casting defects with high melting point metals, and low thermal conductivity, which can lead to unsound casting. It also cannot be used with certain basic metals because it will chemically interact with the metal, forming surface defects. Finally, it releases silica particulates during the pour, risking silicosis in foundry workers.

Olivine sand

Olivine is a mixture of silicates of iron and magnesium from the mineral. Its main advantage is that it is free from silica, therefore it can be used with basic metals, such as manganese steels. Other advantages include a low thermal expansion, high thermal conductivity, and high fusion point. Finally, it is safer to use than silica, therefore it is popular in Europe.

Chromite sand

Chromite sand is a solid solution of spinels. Its advantages are a low percentage of silica, a very high fusion point (1,850 °C (3,360 °F)), and a very high thermal conductivity. Its disadvantage is its costliness; therefore it's only used with expensive alloy steel casting and to make cores.[17]

Zircon sand

Zircon sand is a compound of approximately two-thirds zircon oxide (Zr_2O) and one-third silica. It has the highest fusion point of all the base sands at 2,600 °C (4,710 °F), a very low thermal expansion, and a high thermal conductivity. Because of these good properties it is commonly used when casting alloy steels and other expensive alloys. It is also used as a mold wash (a coating applied to the molding cavity) to improve surface finish. However, it is expensive and not readily available.

Chamotte sand

Chamotte is made by calcining fire clay (Al_2O_3-SiO_2) above 1,100 °C (2,010 °F). Its fusion point is 1,750 °C (3,180 °F) and has low thermal expansion. It is the second cheapest sand, however it is still twice as expensive as silica. Its disadvantages are very coarse grains, which result in a poor surface finish, and it is limited to dry sand molding. Mold washes are used to overcome the surface finish problems. This sand is usually used when casting large steel work pieces.

Other materials

Modern casting production methods can manufacture thin and accurate molds—of a material superficially resembling papier-mâché, such as is used in egg cartons, but that is refractory in nature that are then supported by some means, such as dry sand surrounded by a box, during the casting process. Due to the higher accuracy it is possible to make thinner and hence lighter castings, because extra metal need not be present to allow for variations in the molds. These thin-mold casting methods have been used since the 1960s in the manufacture of cast-iron engine blocks and cylinder heads for automotive applications.

Binders

Binders are added to a base sand to bond the sand particles together (i.e. it is the glue that holds the mold together).

Clay and water

A mixture of clay and water is the most commonly used binder. There are two types of clay commonly used: bentonite and kaolinite, with the former being the most common.

Oil

Oils, such as linseed oil, other vegetable oils and marine oils, used to be used as a binder, however due to their increasing cost, they have been mostly phased out. The oil also required careful baking at 100 to 200 °C (212 to 392 °F) to cure (if overheated, the oil becomes brittle, wasting the mold).

Resin

Resin binders are natural or synthetic high melting point gums. The two common types used are urea formaldehyde (UF) and phenol formaldehyde (PF) resins. PF resins have a higher heat resistance than UF resins and cost less. There are also cold-set resins, which use a catalyst instead of a heat to cure the binder. Resin binders are quite popular because different properties can be achieved by mixing with various additives. Other advantages include good collapsibility, low gassing, and they leave a good surface finish on the casting. MDI (Methylene Diphenyl Isocyanate) is also a commonly used binder resin in the foundry core process.

Sodium silicate

Sodium silicate [Na_2SiO_3 or $(Na_2O)(SiO_2)$] is a high strength binder used with silica molding sand. To cure the binder, carbon dioxide gas is used, which creates the following reaction:

The advantage to this binder is that it can be used at room temperature and is fast. The disadvantage is that its high strength leads to shakeout difficulties and possibly hot tears in the casting.

Additives

Additives are added to the molding components to improve: surface finish, dry strength, refractoriness, and "cushioning properties".

Up to 5% of reducing agents, such as coal powder, pitch, creosote, and fuel oil, may be added to the molding material to prevent wetting (prevention of liquid metal sticking to sand particles, thus leaving them on the casting surface), improve surface finish, decrease metal penetration, and burn-on defects. Up to 3% of "cushioning material", such as wood flour, saw dust, powdered husks, peat, and straw, can be added to reduce scabbing, hot tear, and hot crack casting defects when casting high temperature metals. These materials are beneficial because burn-off when the metal is poured creates tiny voids in the mold, allowing the sand particles to expand. They also increase collapsibility and reduce shakeout time.[21]

Up to 2% of cereal binders, such as dextrin, starch, sulphite lye, and molasses, can be used to increase dry strength (the strength of the mold after curing) and improve surface finish. Cereal binders also improve collapsibility and reduce shakeout time because they burn off when the metal is poured. The disadvantage to cereal binders is that they are expensive.[21]

Up to 2% of iron oxide powder can be used to prevent mold cracking and metal penetration, essentially improving refractoriness. Silica flour (fine silica) and zircon flour also improve refractoriness, especially in ferrous castings. The disadvantages to these additives is that they greatly reduce permeability.

4. STEPS INVOLVED IN CASTING PROCESS

- Pattern Making
- Sand mixing & preparation
- Core making
- Melting
- Pouring
- Finishing
- Testing
- Heat Treatment
- Re-testing

Pattern making

Patternmaking It's the science of making an original pattern or form which will be used to make a mold. The mold will allow for mass production of the original pattern by poring some sort of casting material such as metal, resin, plaster, or other material. A pattern can be both simple and complex for e.g an object that has detail on only one side like a plaque, or with detail on all sides, like a machine part. Factors that go into making a pattern includes durability, shrinkage, machine allowances, draft and undercuts.

In sand casting, the pattern is usually of wood, whereas it may be of metal or other materials in pressure or centrifugal casting. The patternmaker also decides where the sprues and risers (inlet and outlet for molten material) will be placed with respect to the pattern. Objects which have holes or depressions in some of its parts are taken care of by inserting cores. Sometimes patterns may also contain chills, solid pieces of the final material, to enable rapid cooling, resulting in quench hardening in the neighborhood of the chill.

Core making

Cores are placed into a mould cavity to form the interior surfaces of castings. Thus the void space is filled with molten metal and eventually becomes the casting. MouldingMoulding is nothing but the mould preparation activities for receiving molten metal. Moulding usually involves: (i) preparing the consolidated sand mould around a pattern held within a supporting metal frame, (ii) removing the pattern to leave the mould cavity with cores. Mould cavity is the primary cavity. The mould cavity contains the liquid metal and it acts as a negative of the desired product. The mould also contains secondary cavities for pouring and channeling

Melting and Pouring

The preparation of molten metal for casting is referred to simply as melting. The molten metal is transferred to the pouring area where the moulds are filled. Cleaning involves removal of sand, scale, and excess metal from the casting. Burned-on sand and scale are removed to improve the surface appearance of the casting. Excess metal, in the form of fins, wires, parting line fins, and gates, is removed. Inspection of the casting for defects and general quality is performed.

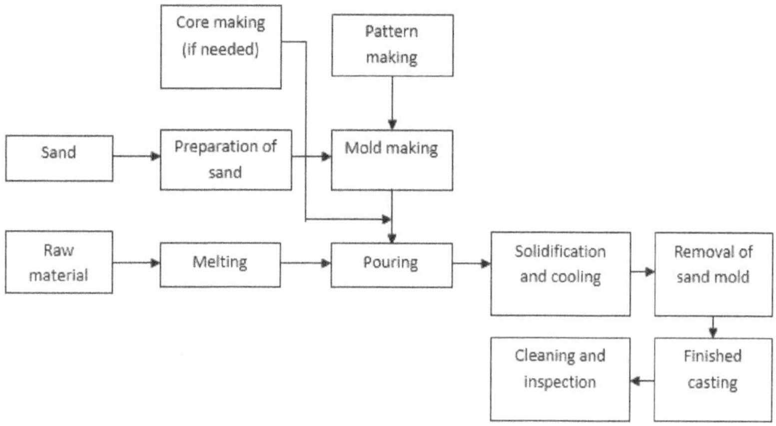

Fig.1 basic steps in making sand castings

Making a simple sand mould

1) The drag flask is placed on the board
2) Dry facing sand is sprinkled over the board
3) Drag half of the pattern is located on the mould board. Dry facing sand will provide a non-sticky layer.
4) Molding sand is then poured in to cover the pattern with the fingers and then the drag is filled completely
5) Sand is then tightly packed in the drag by means of hand rammers. Peen hammers (used first close to drag pattern) and butt hammers (used for surface ramming) are used.
6) The ramming must be proper i.e. it must neither be too hard or soft. Too soft ramming will generate weak mould and imprint of the pattern will not be good. Too hard ramming will not allow gases/air to escape and hence bubbles are created .Vent holes are made in the drag to the full depth of the flask as well as to the pattern to facilitate the removal of gases during pouring and solidification.
7) The finished drag flask is now made upside down exposing the pattern.
8) Cope half of the pattern is then placed on the drag pattern using locating pins. The cope flask is also located with the help of pins. The dry parting sand is sprinkled all over the drag surface and on the pattern.
9) A sprue pin for making the sprue passage is located at some distance from the pattern edge. Riser pin is placed at an appropriate place.
10) Filling, ramming and venting of the cope is done in the same manner
11) The sprue and riser are removed and a pouring basin is made at the top to pour the liquid metal.
12) Pattern from the cope and drag is removed.

13) Runners and gates are made by cutting the parting surface with a gate cutter. A gate cutter is a piece of sheet metal bent to the desired radius.
14) The core for making a central hole is now placed into the mould cavity in the drag. Rests in core prints.
15) Mould is now assembled and ready for pouring.

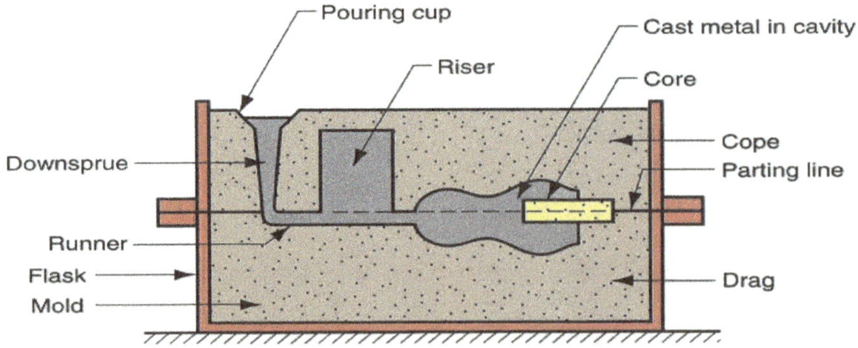

Fig.2 Making sandmould casting

Advantages of Casting Process

- any intricate shape & shape can be cast, both internal & external
- any metal or alloy that can be heated to the liquid state can be cast
- Is the easiest and quickest way from drawing to the production
- Some methods are highly suitable for mass production
- Some operations capable of producing parts to net shape require no further machining
- rapid cooling provides small grain size and good strength to casting

Limitations of Casting Process

- Limitation on mechanical properties
- Poor dimensional accuracy and surface finish
- Removal of pattern of the thin & small parts is very difficult
- Safety hazards to humans & environmental problems
- Internal defects cannot be completely eliminated

Applications of casting process:

(a) Automobile engine blocks, cylinder blocks of automobile
(b) Airplane engines, pistons' and piston rings,
(c) Machine tool beds and frames, mill rolls,
(d) Water supply and sewer pipes, sanitary fittings and agricultural parts etc.

The six basic steps in making sand castings are,

(i) Pattern making,
(ii) Core making,

(iii) Moulding,
(iv) Melting and pouring,
(v) Cleaning Pattern making - Pattern: Replica of the part to be cast and is used to prepare the mould cavity.

5. PATTERN

Pattern is replica or model of object which to be created. It is used to make hollow cavity in sand mold in which molten metal is poured and allow solidifying to create object. The size and shape of cast object is highly depends of shape and size of pattern. Mostly pattern are made by aluminum, wood, wax etc. Metal pattern are used for mass production. The pattern making is most critical work in casting because the object is highly depended on it. A good pattern should follow following requirements.

- Low cost and easy to cast.
- Easy to repair.
- It should light in weight which makes it easy to handle.
- It should able to withstand ramming forces without deformation.
- It should does not change its dimensions in presence of moisture.
- It should be easily removable from mould.
- It should have long life without change in its dimensions.

The important pattern types in casting:

The pattern types in casting process depend on the part to be manufactured. The sand casting pattern making will be easy remove from the mold without any damage in the mold product. It is based on design patterns material and more. The different types of pattern as follows,

1. Solid or single piece pattern
2. Split pattern or two piece pattern
3. Multi-piece pattern
4. Cope and drag pattern
5. Match plate pattern
6. Gated pattern
7. Skeleton pattern
8. sweep pattern sand casting
9. Loose piece pattern
10. Segmental pattern
11. Follow board pattern
12. Shell Pattern

Solid pattern in casting process:

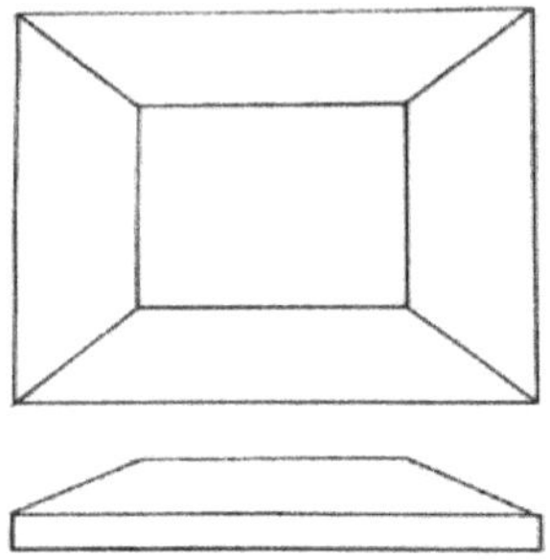

Fig3.Solid Pattern Diagram

The solid pattern types is a most and simple method for simple shape casting. It pattern can make without any sub-part or joint in the mold part. In this type of pattern only produce simple shape and withdrawn for very easily from the mold. The solid pattern placed in the drag position. That is used for make a flat surface like as gear blanks, square blocks and more it solid pattern made depend on design patterns material, shape and more.

Split pattern (or) two piece pattern:

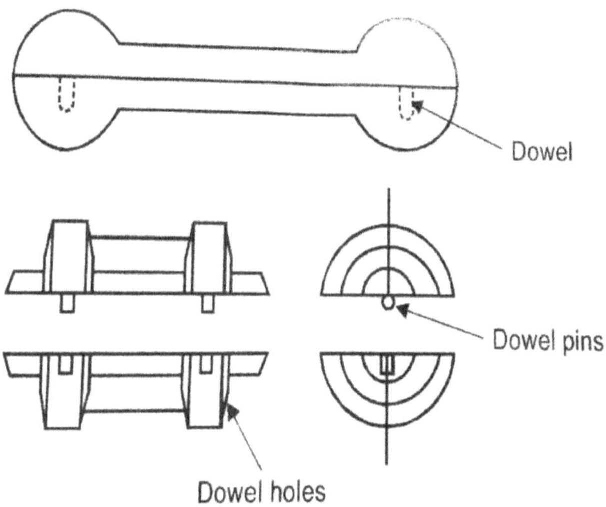

Fig.4 Split Pattern Diagram

When contour of casting manufacture. The sand casting pattern making withdraw from the mold is difficult or when the part depth too high in the casting. The pattern is split into two half part. For one half is contained in the drag and another one in cope. For intricate shape part manufactured using the two or more pattern pieces. The dowel pins used to piece are aligning together. This pattern types is known as split pattern. The split pattern is common method for intricate casting part to produce. The two halves of pattern to align properly by using dowel pin. It placed on the top half of pattern.The split pattern is above figure.

Multi piece pattern:

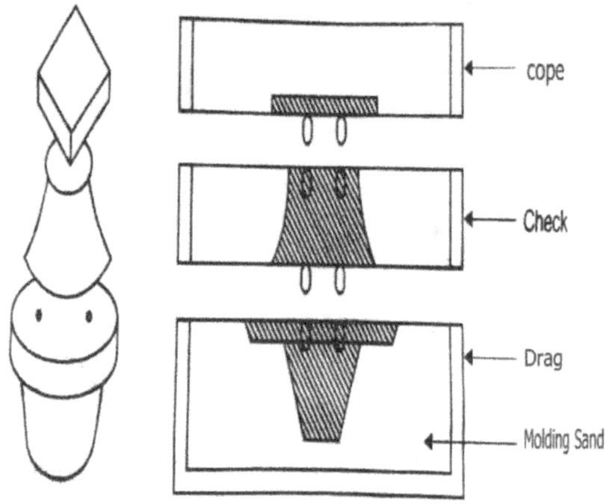

Fig.5 Multi piece pattern Diagram

It is one type of pattern types. When complicated part molded together, to require a sand casting pattern making in more than two parts in order to casting process with easily withdrawal and mold. This pattern contain may be three, or more number of pattern based on the design. For following example of multi piece pattern.

It is having three piece patterns. The top part is cope, bottom is drag and middle part of molding box is called check. They are three pattern will be connected by using dowel pins and molding box clamped by using clamp.

Cope and drag type of pattern:

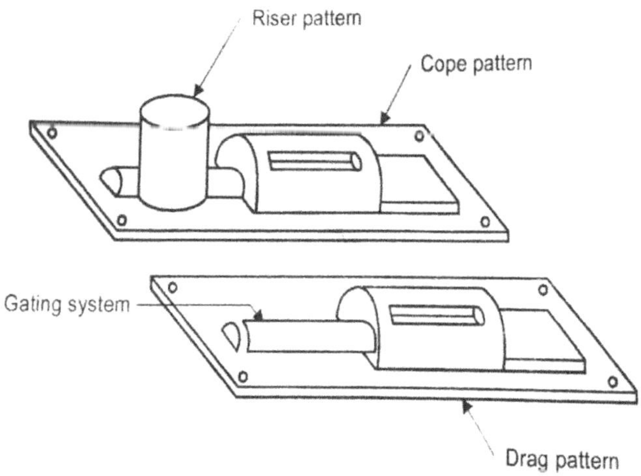

Fig.6 Cope and Drag Pattern Types Diagram

In this pattern types used to provide complicate product. The more complicated part to be made, the complete sand casting pattern making become too heavy to be handled by a single operator. In this Cope and drag type of pattern made in two parts, which separately molded in different molding box. After molding process completed to form of complete mold cavity. When the one part is drag and another one is cope. It is Called cope and drag. It is different from the split pattern because of drag and cope pattern both are molded separately in the assembled position. The cope and drag type of pattern as above diagram.

Match plate type pattern:

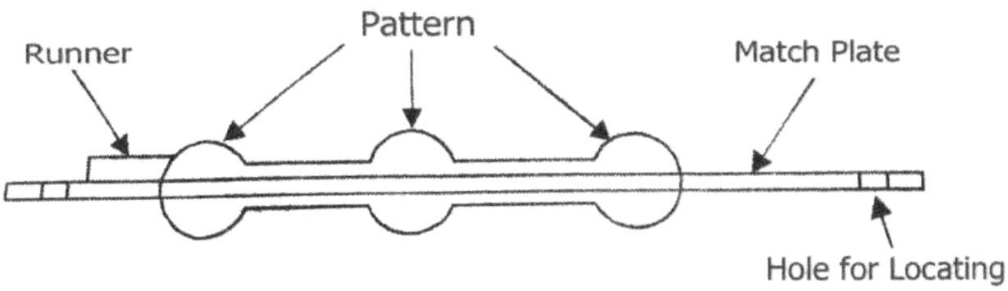

Fig.7 Match Plate Pattern Diagram

The match plate pattern types is having two parts, one for one side and another one for another side of pattern. It is called match plate pattern. The sand casting pattern making in two pieces. It also having gates and runner attached with pattern. The molding process completed after that match plate removed together, the gating is obtained for joining the cope and drag. It pattern is mainly used for casting of metal, usually aluminum are machined in this method with light weight and machinability. It should be possible for mass production of small casting with high dimensional accuracy. They are also used for machine molding. The cost will be high of molding but it is easily compensated by high rate of production and more accuracy.

Gated pattern:

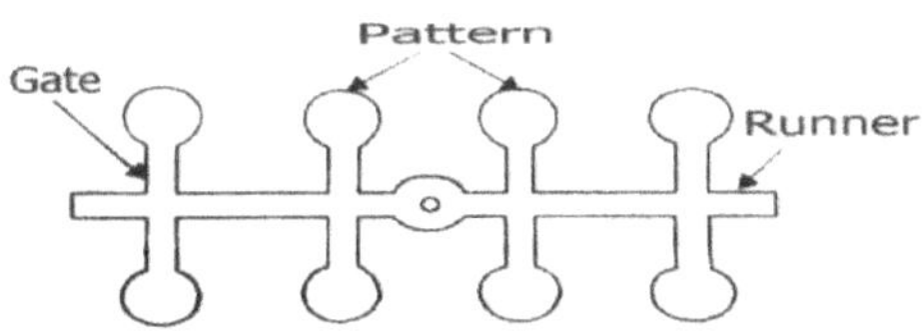

Fig.8 Gated pattern Diagram

To make multiple parts with in single mold and single pattern for all the part cavity of mold. The multi cavity mold is prepared a single sand mold carries a multiple number of cavities. The gates are used to connect the pattern each other. The suitable gates or channels are provided for feeding the molten metal into cavity. All the cavity are feed by using single runner. It mainly consider for low molding time and uniformly feeding of molten metal. It used for mass production of small casting.

Skeleton pattern:

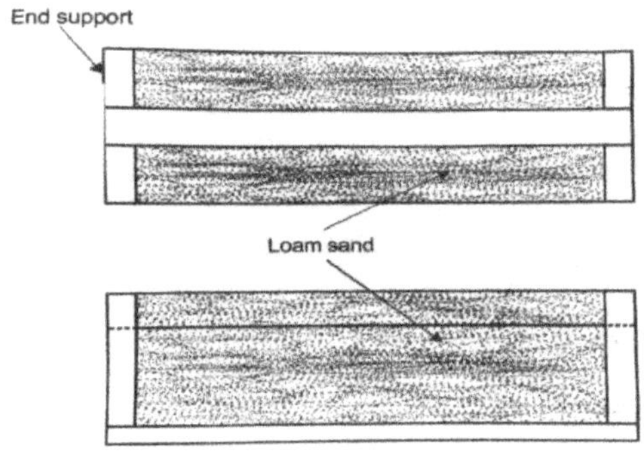

Fig.9 Skeleton pattern Diagram

The casting size is very large but easy to shape and only possible for little number of parts to be made, also not economical with low quantity of large solid pattern of size. In this stage a pattern consists of wooden frame and strips is made. It is called skeleton pattern. The mold is filled properly. The surplus sand is removed together by means of stickler.

Sweep pattern sand casting Method:

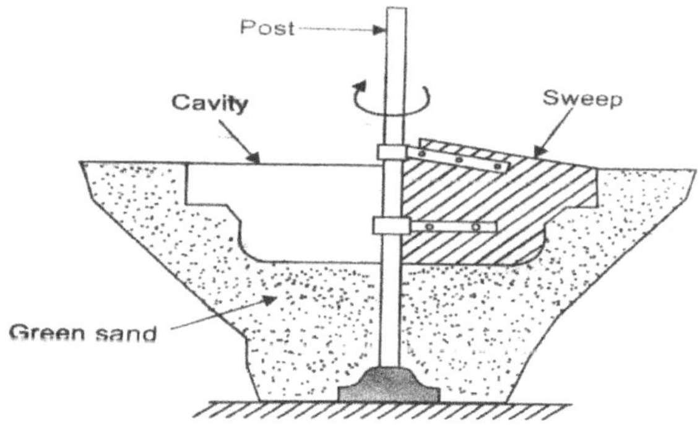

Fig.10 Sweep pattern sand casting Methods

The plane is rotated about an axis with 360^0 called symmetry or sweep. In this sweep pattern sand casting prepare the sand mold by using this method. It is called sweep pattern. It could be economical to save money and make with full pattern because of symmetry. In this pattern used for prepare the mold of large symmetrical casting by mean of circular cross section.

The sweep patterns sand casting consists of base are placed on the sand mass, vertical spindle and wooden template is called sweep. The outer ends of the sweep having corresponding to the shape of require casting. The cavity formed together, from sweep is rotated about the axis. The sweep and spindle is removed from the cavity for leaving the base in the sand. The removal of spindle, in this cause to provide hole and it is patched up by filling the sand. It sweep pattern sand casting Method mainly for Circular solid Part make.

Segmental pattern:

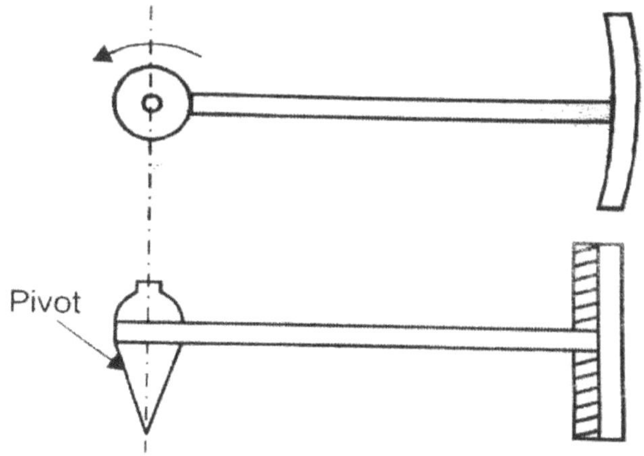

Fig.11 Segmental Pattern Diagram

The segmental pattern is used to prepare the mold of larger circular casting to avoid the use of solid pattern of exact size. It is similar to sweep pattern, but the difference from Sweep pattern, the sweep pattern is give a continuous revolve motion to generate the part, the segmental pattern itself and mold is prepared. In this segmental pattern construction should be save the material for pattern make and easy carried.

The segmental pattern is mounted on the central pivot and mold in one position for after prepare of mold the segment is moved for next position. That is repeat together the complete mold is done.

Loose piece pattern:

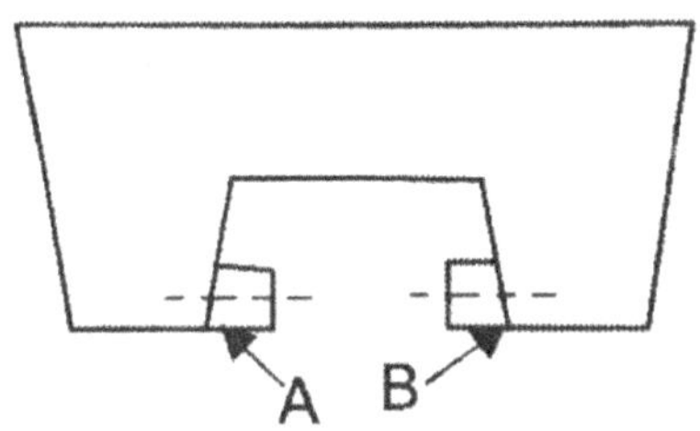

Fig.12 Loose Piece Pattern

A single piece are made to have loose piece in easy to allow withdrawal from the mold the molding process are completed, after the main pattern is withdrawn leaving from that piece in the sand. After the withdrawal of piece from mold, it cavity separately formed by the pattern. It loose piece pattern is highly skilled job and expensive.

Follow board type pattern:

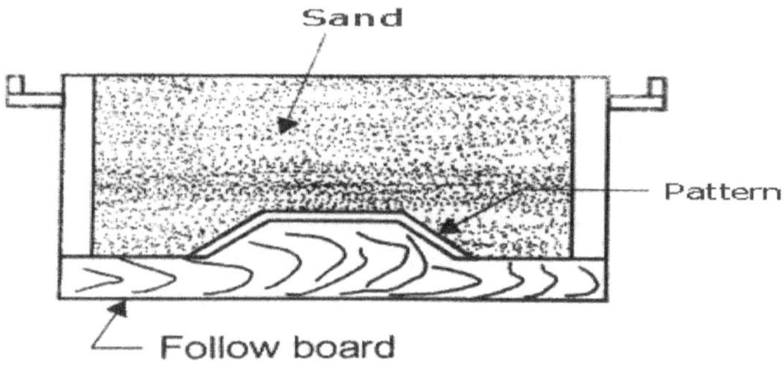

Fig.13Follow Board Pattern

In casting process some portions are structurally weak. It is not supported properly and may be break under the force of ramming. In this stage the special pattern to allow the mold may be such as wooden material.

Shell pattern:

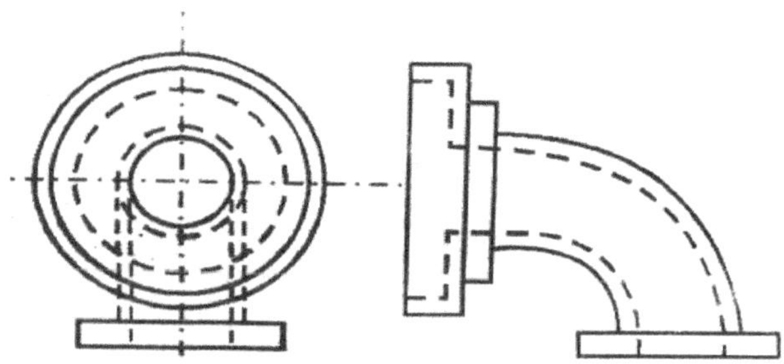

Fig.14 Shell Pattern Diagram

It shell pattern is used to molding of hollow shape product wit curved or straight. It means of pipe work done it process. It pattern usually made of metal. That pattern parted along with the center line and both halves are doweled.

6. PATTERN MATERIALS

The common materials of which the patterns are made are the following:

1) Wood:

It is the most common material used for pattern making because of the following

Advantages:

(i) It is cheap and available in abundance.
(ii) It can be easily shaped into different forms and intricate designs.
(iii) Its manipulation is easy because of lightness in weight.
(iv) Good surface finish can be easily obtained by only planning and sanding.

(v) It can be preserved for a fairly long time by applying proper preservatives like shellac varnish.

Disadvantages:

(i) It wears out quickly due to its low resistance to sand abrasion. As such, a wooden pattern cannot stand a long constant use.

(ii) It is very susceptible to moisture, which may lead to its warping or splitting. This needs its careful storing in a dry place and the application of preservatives.

(iii) Its life, owing to the above reasons, is short as compared to other pattern materials. This confines its use to such cases only when a small number of castings are required.

2) Metals:

Metals are used with advantage, as pattern material, only when the number of castings to be made is very high and a closer dimensional accuracy is desired. They have a much longer life than wooden patterns and eliminate the inherent disadvantages of wood to a great extent.
But they also carry the following

Disadvantages:

(i) They are costlier than wood and, therefore, cannot be used with advantage, where a smaller number of castings is to be made.

(ii) For giving different shapes and fine surface finish they need machining. This again adds to their cost.

(iii) Most of them are very heavy and in case of large castings the weight of the pattern always poses a problem in its manipulation.

(iv) A large number of them have a tendency to get rusted.

3) Plaster:

Plaster of Paris or gypsum cement is advantageously used as a pattern material since it can be easily casted into intricate shapes and can be easily worked also. Its expansion can be easily controlled and it carries a very high compression strength. Its specific use is in making small patterns and core boxes involving intricate shapes and closer dimensional control. A marked feature of this cement is that contrary to the action of metals, it expands on being solidified. Thus, if a cement of proper coefficient of expansion is selected, the effect of shrinkage of casting can be automatically neutralized.

4) Plastics:

Plastics are gradually gaining favor as pattern materials due to their following specific characteristics:

1. Lightness in weight.
2. High strength.
3. High resistance to wear.
4. High resistance to corrosion due to moisture.
5. Fine surface finish.
6. Low solid shrinkage.
7. Very reasonable cost.

The plastics used as pattern materials are thermosetting resins. Phenolic resin plastic and foam plastic suit best for this purpose. For making the pattern, first the moulds are made, usually from plaster of Paris. The resin is then poured into these moulds and the two heated. At a specific temperature, the resin solidifies to give the plastic pattern.

5) Wax:

Wax patterns are exclusively used in investment casting. For this a die or metal mould is made in two halves into which the heated wax is poured. The die is kept cool by circulating water around it. As the wax sets on cooling, the die parts are separated and the wax pattern taken out

7. PATTERN ALLOWANCES

A pattern is always made larger than the required size of the casting considering the various allowances. These are the allowances which are usually provided in a pattern.

1: Shrinkage or contraction allowance:

The various metals used for casting contract after solidification in the mould. Since the contraction is different for different materials, therefore it will also differ with the form or type of metal.

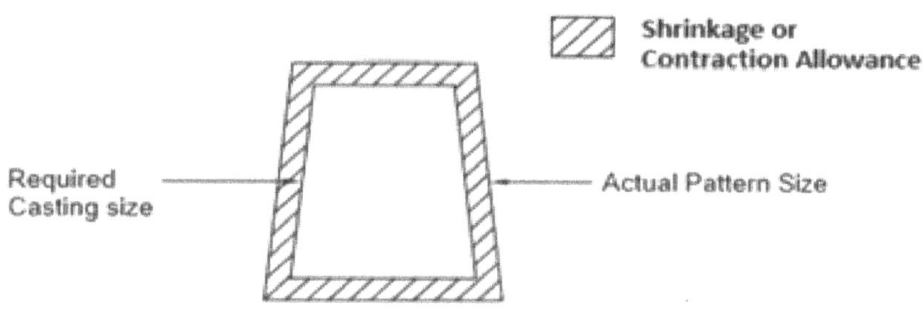

Fig.15 Shrinkage or contraction allowance

2: Draft allowance

It is a taper which is given to all the vertical walls of the pattern for easy and clean withdraw of the pattern from the sand without damaging the mould cavity. It may be expressed in millimeters on a side or in degrees. The amount of taper varies with the type of patterns. The wooden patterns require more taper than metal patterns because of the greater frictional resistance of the wooden surfaces.

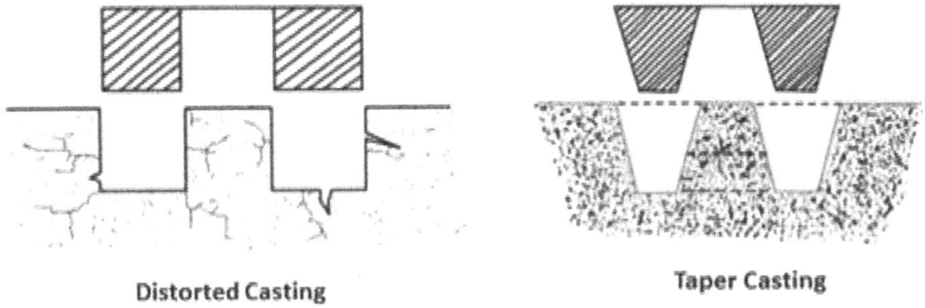

Fig.16Draft allowance

3: Finish or machining allowance

The allowance is provided on the pattern if the casting is to be machined. This allowance is given in addition to shrinkage allowance. The amount of this allowance varies from 1.6 to 12.5 mm which depends upon the type of the casting metal, size and the shape of the casting. The ferrous metals require more machining allowance than non ferrous metals.

4: Distortion or camber allowance

This allowance is provided on patterns used for casting of such design in which the contraction is not uniform throughout.

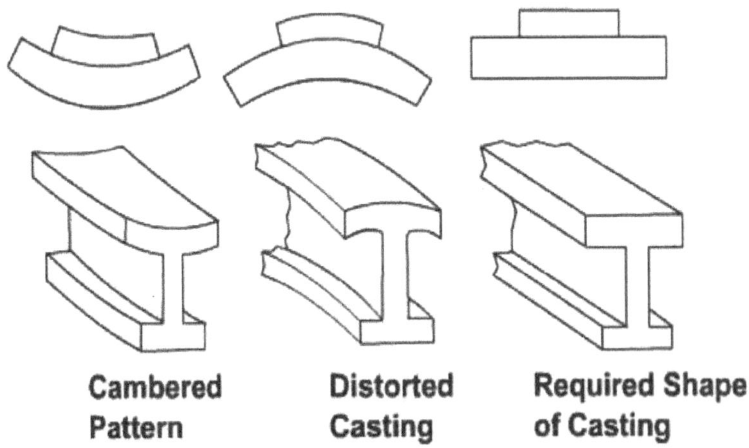

Fig.17Distortion or camber allowance

5: Rapping or shaking allowance

This allowance is provided in the pattern to compensate for the rapping of mould because the pattern is to be rapped before removing it from the mould.

8. TYPES OF MOLDING SAND

According to the use, molding sand may be classified as below:

1. Green Sand:

The green sand is the natural sand containing sufficient moisture in it. It is mixture of silica and 15 to 30% clay with about 8% water. Clay and water act as a bonding material to give strength. Molds made from this sand are known as green sand mould.

The green sand is used only for simple and rough casting products. It is used for both ferrous and non-ferrous metals.

2. Dry Sand:

When the moisture is removed from green sand, it is known as dry sand. The mould produced by dry sand has greater strength, rigidity and thermal stability. This sand is used for large and heavy castings.

3. Loam Sand:

Loam sand is a mixture of 50 percent sand and 50 percent clay. Water is added in sufficient amount. It is used for large and heavy moulds e.g., turbine parts, hoppers etc.

4. Facing Sand:

A sand used for facing of the mould is known as facing sand. It consists of silica sand and clay, without addition of used sand. It is used directly next to the surface of the pattern. Facing sand comes in direct contact with the hot molten metal; therefore it must have high refractoriness and strength. It has very fine grains.

5. Parting Sand:

A pure silica sand employed on the faces of the pattern before molding is known as parting sand. When the pattern is withdrawn from the mould, the molding sand sticks to it.

To avoid sticking, parting sand is sprinkled on the pattern before it is embedded in the molding sand. Parting sand is also sprinkled on the contact surface of cope, drag and cheek.

6. Backing or Floor Sand:

The backing sand is old and repeatedly used sand of black color. It is used to back up the facing sand and to fill the whole volume of the box. This sand is accumulated on the floor after casting and hence also known as floor sand.

7. System Sand:

The sand employed in mechanical heavy castings and has high strength, permeability and refractoriness, is known as system sand. It is used for machine molding to fill the whole flask. In machine molding no facing sand is used. The system sand is cleaned and has special additives.

8. Core Sand:

A sand used for making cores is known as core sand. It is silica sand mixed with core oil (linseed oil, resin, mineral oil) and other binding materials (dextrine, corn flour, sodium silicate). It has remarkable compressive strength.

9. Molasses Sand:

A sand which carries molasses as a binder is known as molasses sand. It is used for core making and small castings of intricate shapes.

9. PROPERTIES OF MOLDING SAND

1. Porosity:

Porosity also known as permeability is the most important property of the molding sand. It is the ability of the molding sand to allow gasses to pass through. Gasses and steam are generated during the pouring of molten metal into the sand cavity. This property depends not only on the shape and size of the particles of the sand but also on the amount of the clay, binding material, and moisture contents in the mixture.

2. Cohesiveness:

Cohesiveness is the property of sand to hold its particles together. It may be defined as the strength of the molding sand. This property plays a vital role in retaining intricate shapes of the mould.

Insufficient strength may lead to a collapse in the mould particles during handling, turning over, or closing. Clay and bentonite improves the cohesiveness.

3. Adhesiveness:

Adhesiveness is the property of sand due to which the sand particles sticks to the sides of the molding box. Adhesiveness of sand enables the proper lifting of cope along with the sand.

4. Plasticity:

Plasticity is the property of the molding sand by virtue of which it flows to all corners around the mould when rammed, thus not providing any possibility of left out spaces, and acquires a predetermined shape under ramming pressure.

5. Flow-Ability:

Flow-ability is the ability of molding sand to free flow and fill the recesses and the fine details in the pattern. It varies with moisture content.

6. Collapsibility:

Collapsibility is the property of sand due to which the sand mould collapse automatically after the solidification of the casting. The mould should disintegrate into small particles of molding sand with minimum force after the casting is removed from it.

7. Refractoriness

Refractoriness is the property of sand to withstand high temperature of molten metal without fusion or soften.

Molding sands with poor refractoriness may burn when the molten metal is poured into the mould. Usually, sand moulds should be able to withstand up to 1650°C.

Melting

Melting is performed in a furnace. Virgin material, external scrap, internal scrap, and alloying elements are used to charge the furnace. Virgin material refers to commercially pure forms of the primary metal used to form a particular alloy. Alloying elements are either pure forms of an alloying element, like electrolytic nickel, or alloys of limited composition, such as ferroalloys or master alloys. External scrap is material from other forming processes such as punching, forging, or machining. Internal scrap consists of gates, risers, defective castings, and other extraneous metal oddments produced within the facility.

The process includes melting the charge, refining the melt, adjusting the melt chemistry and tapping into a transport vessel. Refining is done to remove harmful gases and elements from the molten metal to avoid casting defects. Material is added during the melting process to bring the final chemistry within a specific range specified by industry and/or internal standards. Certain fluxes may be used to separate the metal from slag and/or dross and degassers are used to remove dissolved gas from metals that readily dissolve certain gasses. During the tap, final chemistry adjustments are made.

10. FURNACE

Several specialized furnaces are used to heat the metal. Furnaces are refractory-lined vessels that contain the material to be melted and provide the energy to melt it. Modern furnace types include Electric Arc Furnaces (EAF), induction furnaces, cupolas and crucible furnaces. Furnace choice is dependent on the alloy system quantities produced.

Crucible Furnaces

Crucible furnaces are one of the oldest and simplest types of melting unit used in the foundry. The furnaces use a refractory crucible which contains the metal charge. The charge is heated via conduction of heat through the walls of the crucible. The heating fuel is typically coke, oil, gas or electricity. Crucible melting is commonly used where small batches of low melting point alloy are required. The capital outlay of these furnaces makes them attractive to small non-ferrous foundries.

Crucible furnaces are typically classified according to the method of removing the metal from the crucible:

Tilting furnace, in which the molten metal is transferred to the mould or ladle by mechanically tilting the crucible and furnace body. Lift-out furnace, in which the crucible and molten metal are removed from the furnace body for direct pouring into the mould. Bale-out furnace, in which the metal is ladled from the crucible to the mould.

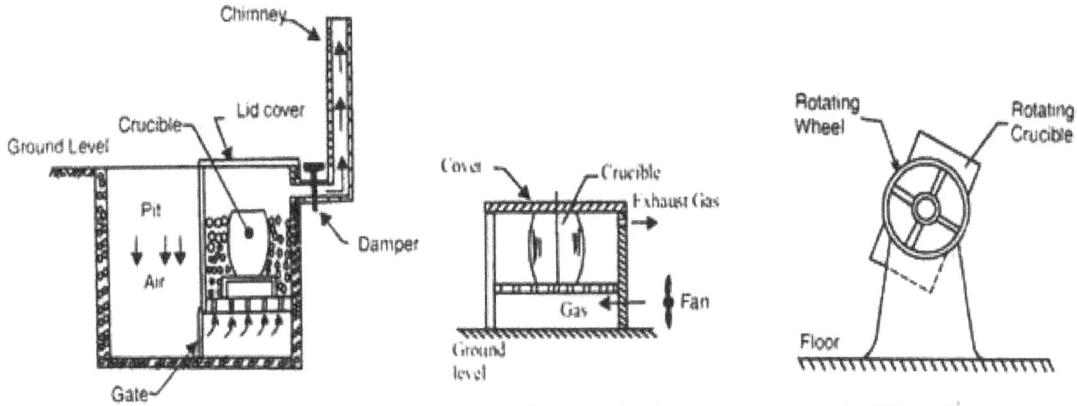

(*a*) Pit furnace (stationary type). (*b*) Bale-out furnance. (stationary type). (*c*) Tilting furnace.

Fig.18Crucible Furnaces

Introduction to Cupola Furnace:

The cupola is a most widely used foundry furnace for melting ferrous metals and alloys. Sometimes, it is also used for melting non-ferrous metals and alloys. Cupola furnace is the cheapest means for converting pig iron or scrap metal into gray cast iron. The fuel used is a good quality low sulphur coke. Anthracite coal or carbon briquettes may also be used.

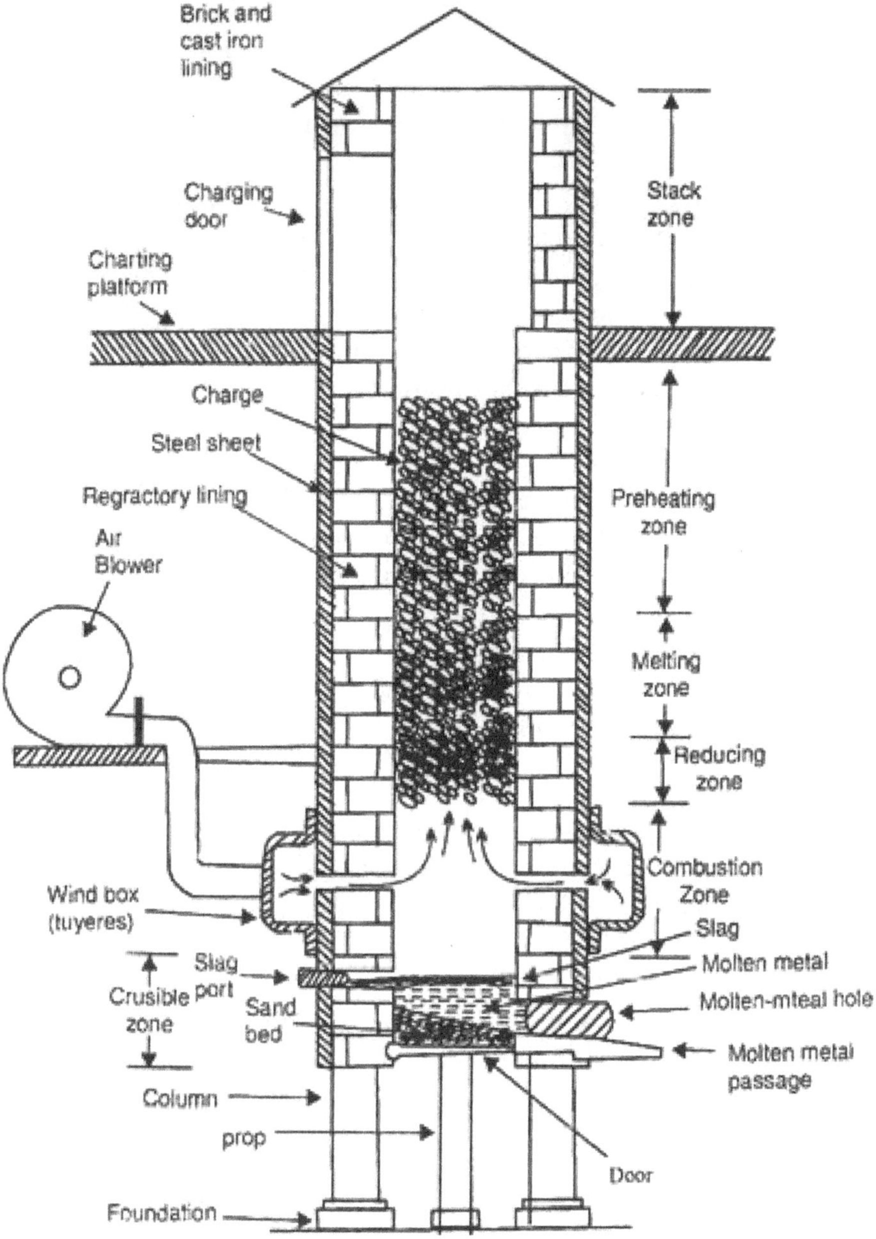

Fig.19 Cupola Furnace

Structure of Cupola Furnace:

The cupola is a shaft type furnace whose height is three to five times its diameter; it is most widely used furnace for producing molten gray cast iron. A sketch of a cupola furnace is shown in Fig.

As seen from the Fig, the main parts of cupola are:

(i) Shell:

The shell is constructed of a steel plate of about 10 mm thick riveted or welded together and that is internally lined with refractory fireclay bricks. The shell diameter varies from 1 to 2 meters with a height of about three to five times the diameter.

(ii) Foundation:

The whole structure is erected on legs or steel columns. A drop door, which is made of one piece, is hinged to a supporting leg. When the cupola is full of charge, a prop support at the bottom door is provided so that door remains close and do not collapse due to the heavy weight of the charge. If the cupola is not in use, the drop door allows for maintenance and repair work of the furnace lining.

(iii) Charging Door:

Towards the top of the furnace there is an opening called charging door is situated. The charging door is used for feeding the charge containing metal, coke and flux, into the furnace. It is situated at the height of about 3 to 6 m above the tuyers.

(iv) Charging Platform:

The charging platform is made of robust mild steel rods and plates. Usually, it is surrounding the cupola at the level of about 0.3 m below the bottom of the charging door.

(v) Air Blower:

An air blower is connected to the wind box by means of blast pipe. It supplies the air to the wind box. A valve is provided in the blast pipe to control the flow of air. The blast pressure varies from 250 kg/m2 to 1050 kg/m2.

(vi) Tuyeres:

Air, which is needed for the combustion is blown through the tuyeres located about 36 inches (0.9 m) above the bottom of the furnace. Total area of the tuyeres should be 1/5 to 1/6 of the cross-section area of the cupola inside the lining at tuyere level.

(vii) Volume Meter:

The volume meter is installed in a cupola furnace to know the volume of air passing. The amount of air needed to melt one tone of iron depends upon the quality and quantity of coke and coke iron ratio.

(viii) Tap Hole (Molten Metal Hole):

Slightly above the bottom and in the front there are a tap hole to allow molten cast iron to be collected.

(ix) Slag Hole:

There is also a slag hole located at the rear and above the level of the tap hole because slag floats on the surface of molten cast iron.

(x) Chimney:

The portion of shell above charging hole is known as chimney. Its height is generally 4 to 6 m. The chimney is provided with a filter screen and a spark arrester. This facilitates a free escape of the waste gases and deflects the sparks and the dust back into the furnace.

Operation of Cupola Furnace:

The operation of cupola furnace consists of following steps:

(i) Preparation of Cupola:

A newly built cupola should be thoroughly dried before firing. Any slags around the tuyeres from previous run are cleaned. Any broken bricks are repaired with a mixture of silica sand and fire clay. A layer of refractory material is applied over the brunt area over the fire brick lining.

A bed of molding sand is then rammed on the bottom to a thickness of about 6 inches (15 cm) or more, sloping towards the tap hole to ensure better flow of molten metal. A slag hole opening of about 30 to 35 mm diameter and a tap hole of about 25 mm diameter is provided.

(ii) Firing of Cupola:

A fire of wood is ignited on the sand bottom, when the wood burns well; coke is dumped on the bed well from top. Make sure that the coke begins to burn too. A bed of coke about 40 inches thick is next placed on the sand i.e., slightly above the tuyeres.

The air blast is turned on at a lower blowing rate than normal for igniting the coke. A measuring rod is used which indicates the height of coke bed. Firing is done about 3 hours before the molten metal required.

(iii) Charging the Cupola:

Next, the charge is fed into the cupola through the charging door. Many factors, such as the charge composition, affect the final structure of the gray cast iron obtained. The charge is composed of 25% pig iron, 50% gray cast iron scrap, 10% steel scrap, 12% coke as fuel, and 3% limestone as flux.

These constituents form alternate layers of coke, limestone and metal. Besides limestone, fluorspar and soda ash are also used as flux material - The function of flux is to remove the impurities in the iron and protect the iron from oxidation.

(iv) Soaking of Iron:

After charging the furnace fully, it is allowed to remain as such for about 1—1.5 hr. During this stage charge slowly gets heated up because the air blast is kept shut this time and due to this the iron gets soaked.

(v) Starting the Air Blast:

The air blast is opened at the end of the soaking period. The top opening is kept closed till the metal melts and sufficient metal is collected. As melting proceeds, the contents of the charge move gradually downwards. The rate of charging must be equal to the rate of melting so that the furnace is kept full throughout the heat.

(vi) Closing the Cupola:

When no more melting is required, the feeding of charge and air blast is stopped. The prop is removed, so that the bottom plate swings to open. The slag deposited is removed. The cupola can run continuously as are blast furnace, but in practice it may works when required. The melting period does not exceed 4 hours in most of the foundries. But, it can be operated continuously for 10 hours or more.

Zones of Cupola Furnace:

The cupola furnace is divided in a number of zones where a number of chemical reactions take place.

The following are the six important zones:

(i) Well or Crucible Zone:

It is the zone between top of the sand bed and bottom of the tuyeres. Molten metal collected in this zone.

(ii) Combination Zone:

It is the zone between the top of the tuyeres and a theoretical level above it. It is also known as oxidizing zone.

Here, the combustion actually done, consuming all the oxygen from the air blast and generates huge amount of heat. The temperature range for this zone is about 1500°C to 1850°C. The heat produced in this zone is sufficient to meet the requirements of other zones of cupola.

(iii) Reducing Zone:

It is the zone between the top of the combustion zone and the top level of the coke bed. It is also known as protective zone.

The Co_2 flowing upward through this zone reacts with hot coke and Co, is reduced to Co. Due to this reaction, the temperature gets reduced to about 1200°C. This zone protects the charge against oxidation as it has reducing atmosphere in it.

(iv) Melting Zone:

It is the zone between the first layer of metal charge and above the reducing zone. It is between 300 to 900 mm above the bed charge. The solid metal charge changes to molten state picks up sufficient carbon in this zone. The temperature attainable in this zone is in the range of 1600°C to 1700°C.

(v) Preheating Zone:

It is the zone from above the melting zone to the bottom level of the charging door. Charging materials are fed in this zone. The charge is preheated to about 1093°C before they settle downwards to enter the melting zone. It is also known as charging zone.

(vi) Stack Zone:

It is the empty portion of this furnace, which extends from above the charging zone to the top of the furnace. It carries the hot gases generated within the furnace to the atmosphere.

Capacity of Cupola Furnace:

The capacity of cupola is defined in terms of tones of liquid metal obtained per hour of heat. It depends upon the dimensions of cupola, the efficiency of combustion, combustion rate, constituents of charging, etc. The output of cupola can be increased by oxygen enrichment of air blast and by better heat utilization of hot outgoing gases to preheat the furnace to about 180 to 270°C.

Advantages of Cupola Furnace:

(i) It is simple in construction and operation.
(ii) Low cast of construction, operation and maintenance.
(iii) It has a continuous and fast rate of production.
(iv) It does not require very skilled operators.
(v) It requires small floor area as compared to other furnaces.
(vi) Composition of melt can be controlled.

Limitations of Cupola Furnace:

(i) Temperature control is difficult to maintain.
(ii) Carbon content increases in the iron product due to the heating of coke together with metal.

Some metal elements are converted to their oxides, which are not suitable for casting.

11. CASTING DEFECTS AND CAUSES

Gas porosity occurs when the metal traps gas (most often nitrogen, oxygen or hydrogen) during casting.

When the casting cools and solidifies, bubbles form because the solid form of the metal cannot hold as much gas as the liquid form. These bubbles appear on a casting as rounded, circular cavities or holes.

There are three types of casting defects related to gas porosity:

1. Pinholes

Pinholes, also sometimes referred to as porosities, are very tiny holes (about 2 mm) usually found in the cope (upper) part of the mold, in poorly vented pockets.

They usually appear in large numbers together, either at the surface or just below the surface of the casting. They are always visible to the naked eye and don't require equipment to identify.

2. Subsurface blowhole

Blowholes, or simply blows, are larger cavities than pinholes.

A subsurface blowhole appears on the inside of a cast and usually isn't visible until after machining. Subsurface blowholes can be difficult to detect before machining, requiring harmonic, ultrasonic, magnetic or x-ray analysis.

3. Open holes

These blowholes appear on the surface of the cast and are easier to detect than subsurface blowholes.

Causes and prevention of gas porosity

There are several causes of cavity defects.

- Poor venting of mold and cores
- Insufficient drying of mold and cores

Scars are shallow blows that appear on a flat surface, while **blisters** are scars covered with a thin layer of metal.

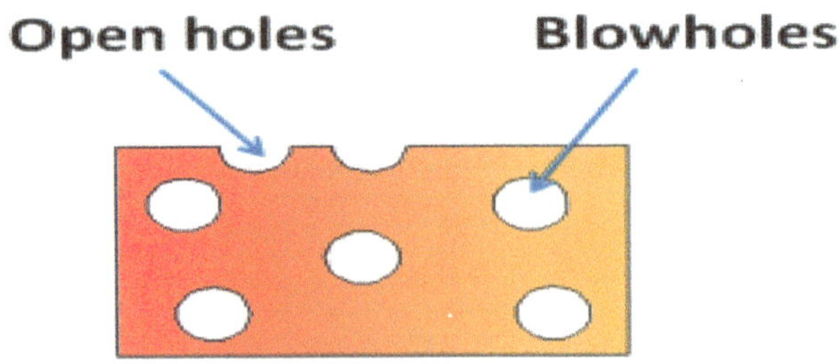

Fig.20 Porosit

- Excessive moisture content of molding sand
- Inadequate gas permeability of molding sand

Potential solutions include:

- Incorporate good fluxing and melting practices: melt metal in a vacuum, in an environment of low-solubility gases or undera flux that prevents contact with the air
- Increase gas permeability of sand: coarser sands have a higher permeability
- Increase permeability of mold and cores. Allow air and gas to escape from the mold cavity
- Dry out molds and cores before use and store dry
- Increase rate of solidification by reducing metal temperature during casting

Shrinkage casting defects and causes

Shrinkage occurs because metals are less dense as a liquid than a solid.

A shrinkage cavity is a depression in a casting which occurs during the solidification process. Shrinkage porosity appears with angular edges, compared to the round surfaces of gas porosity. Cavities might also be paired with dendritic fractures or cracks.

Large shrinkage cavities can undermine the integrity of the casting and may cause it to eventually break under stress.

Shrinkage can result in two types of casting defects.

4. Open shrinkage defects

These are open to the atmosphere. Air compensated as the shrinkage cavity forms.

Pipes are open shrinkage defects that form at the surface and burrow into the casting. **Caved surfaces** are shallow, open shrinkage defects that form across the surface of the casting.

5. Closed shrinkage defects

Also known as **shrinkage porosity**, closed shrinkage defects form within the casting. Macro shrinkage can be viewed with the naked eye, but micro shrinkage cannot.

Closed shrinkage defects usually appear at the top of hot spots, or isolated pools of hot liquid.

Prevent shrinkage cavities by improving casting structure

Alloys always shrink when changing from molten to solid. This is because the density of a casting alloy in the molten state is lower than that in the solid state.Factor a shrinkage allowance into the pattern design before casting.

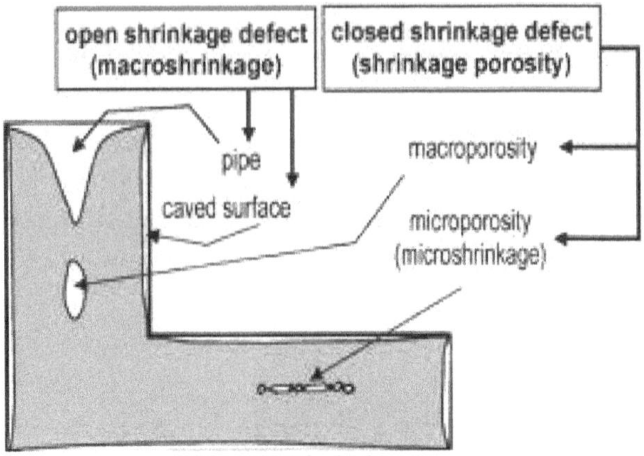

Fig.20 Shrinkage

To prevent shrinkage casting defects by improving the overall casting structure:

- Design a running (gate) system with risers that ensure a continuous flow of molten metal
- Increase local heat dissipation by inserting internal chills, cooling ribs or cooling coils
- Reduce casting temperature to limit the total volume deficit

Mold material casting defects and causes

Mold material casting defects are related to the mold material, which is most commonly sand. You and your supplier can typically address these casting defects and causes by modifying the mold.

6. Cuts and washes

Cuts and washes are areas of excess metal. These appear when the molten metal erodes the molding sand.

A cut appears as a low projection along the surface of the drag face, decreasing in height as it extends from one side of the casting to the other.

Causes and prevention of cuts and washes

Cuts and washes can be caused by molten metal flowing at a high velocity, causing too much metal to flow through the gate.

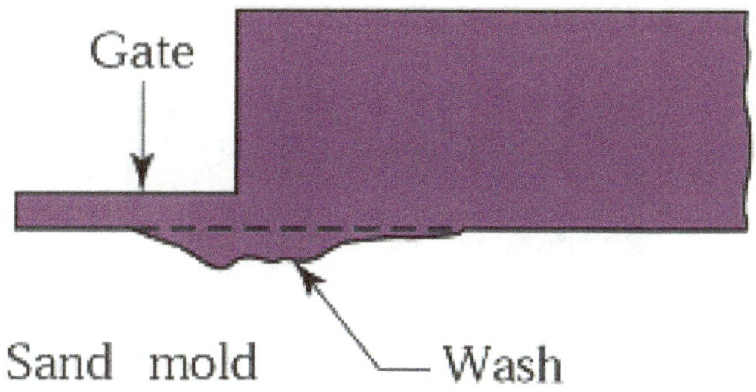

Fig.21 Cuts and washes

To prevent cuts and washes easiest by:

- Designing the gating system properly
- Improving mold and core strength
- Adding more binders to the facing and core sand

7. Fusion

Fusion occurs when sand grains fuse with molten metal. It appears as a thin crust with a brittle, glassy appearance firmly adhered to the casting.

Fig.21 Fusion

Causes and prevention of fusion

Two main factors can cause fusion:

- Low refractoriness of clay or sand
- Too high pouring temperature of molten metal

Refractoriness is the ability of the molding material to resist the temperature of the liquid so it doesn't fuse with the metal. Silica sand has the highest refractoriness.

Improving the refractoriness of the molding material and/or reducing the pouring temperature of the molten metal will help prevent fusion.

8. Run out

Run out is when liquid metal leaks out of the mold, leading to an incomplete or missing casting.

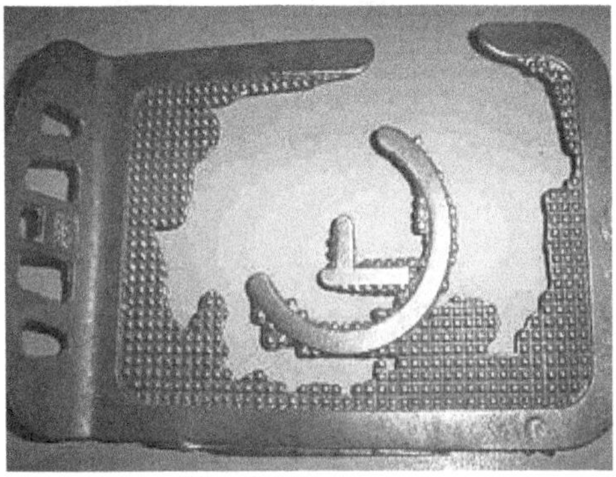

Fig.22 Run out

A faulty mold or flask is responsible for run out.

Prevention of run out and incomplete castings

To prevent this casting defect, design the casting mold with precision. Inspect and replace any defective molds before casting. High temperatures can lead to excess wear and tear of the mold. Use quality raw materials for your mold that can resist high temperatures.

9. Swells

Swells are an enlargement of the casting. Swells typically take on the shape of a slight, smooth bulge on the vertical face of castings.

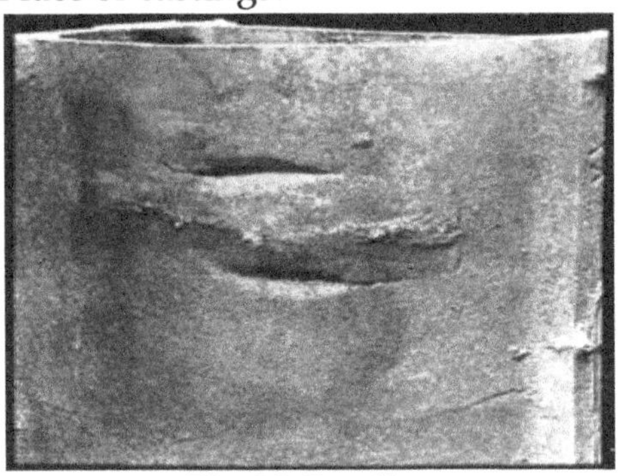

Fig.23 Swells

Causes and prevention of swells

Swell is usually caused by improper or soft ramming of the mold or a low strength mold.

Molds should be built to withstand liquid metal pressure. Otherwise, the mold wall may give way or move back, causing swelling. Using a strong, properly rammed mold prevents swells.

10. Drops

Drops occur when pieces of sand fall into metal casting when it's still liquid. Drops appear as an irregularly shaped projection on the cope (top) surface of a casting.

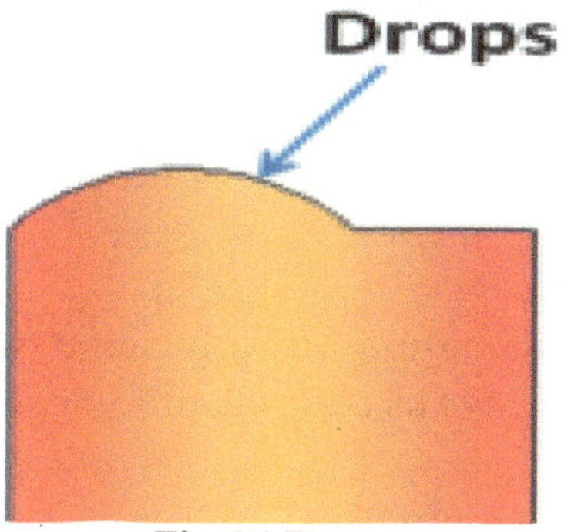

Fig.24 Drops

Causes and prevention of drops

Four potential causes for drops and their preventions include:

- **Low sand strength:** Use sand of a higher strength if this your culprit
- **Soft ramming:** Provide harder ramming
- **Insufficient fluxing of molten metal**: Properly fluxing molten metal removes impurities
- **Insufficient reinforcement of sand projections in the cope**: Reinforce sand projections using nails or gaggers to fix this issue

11. Rat tails, veins and buckles

Fig.25 Rat tails, veins and buckles

Rat tails, or veins, appear as an irregular line or crack on the casting, when the surface of the molding sand buckles up. Rat tails usually occur on the surface of the mold bottom, an area covered with molten material.

Buckles are a more severe form of rat tails.

Causes and prevention of rat tails and buckles

Rat tails and buckles occur when excessive heat of the metal causes the sand to expand. This may be caused by:

- **Poor expansion properties of the sand**: Add combustible additives to sand.
- **A hot pouring temperature**: Reduce pouring temperature of metal.
- **Poor mold design**: Large and flat sections are more prone to rat tails. The mold also should not be too hard, as it must allow for proper expansion.

12. Metal penetration

Metal penetration occurs when liquid metal penetrates gaps in the molding sand. The penetration is visible to the naked eye as a rough and uneven surface finish of the casting.

Causes and prevention of metal penetration

Metal penetration is due to:

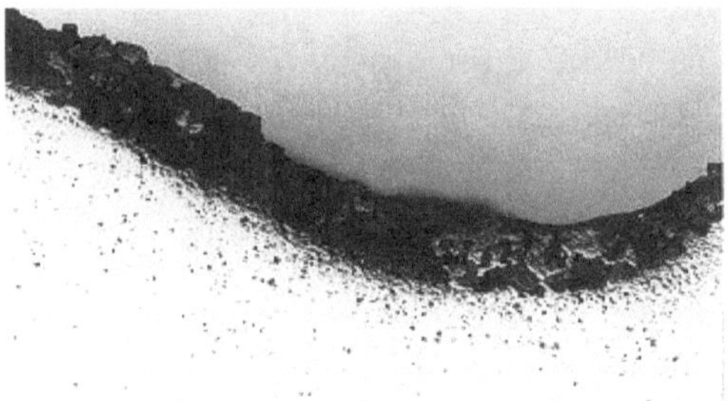

Fig.26 Metal penetration

- Use of sand with low strength and high permeability
- Use of large or coarse sand grain: the coarser the sand grains, the more severe the metal penetration
- Lack of mold wash
- Soft ramming of sand

Prevent metal penetration by fixing these areas. Use high strength, small grain size, low permeability and hard ramming of sand. Ensure a protective barrier against metal penetration by coating the surface of molds with a mold wash.

You can typically remove metal penetration by grinding down the rough surface of the casting.

Metallurgical casting defects and causes

There are two types of metallurgical defects to watch out for.

13. Hot tear/crack

Cracks appear in the form of irregular crevices in a branched pattern.

Fig.27 Cracks

Some cracks are obvious and easily seen, while others can require magnification. Cracks occur as the casting cools, towards the end of solidification.

Causes and prevention of hot tears and cracks

If the solidifying metal does not have sufficient strength to resist tensile forces during solidification, hot tears will appear.

Hot tears are mostly caused by poor mold design. Modifying the mold to improve collapsibility can easily resolve these issues.

14. Hot/hard spots

Hot spots are spots that are harder than the surrounding area. This is because they cooled more quickly than the surrounding material

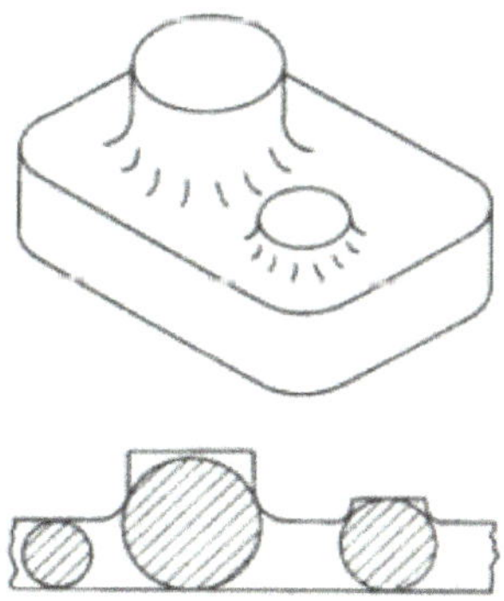

Fig.28 Hot spots

Hard spots can interfere with machining and increase tool wear.

Causes and prevention of hot spots

Hot spots are a direct result of improper cooling practices. There are two potential solutions if hot spots are your problem:

- Start by correcting cooling practice
- Also consider changing the metal's chemical composition

Pouring casting defects and causes

Pouring metal defects arise during the process of pouring metal into the mold. If you have a pouring metal defect, it will fall into one of these categories:

15. Cold shut/lap

Cold shut is a type of surface defect. You'll see a line or crack with a round edge on the casting surface. This defect is visible to the naked eye and often results in rejecting the cast, as it creates a weak spot.

Causes and prevention of cold shut

When molten metal enters the mold from two gates, the streams will meet at a junction

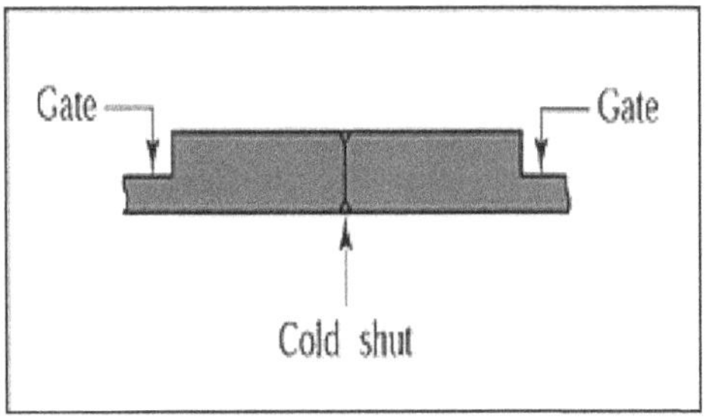

Fig.29 Cold shut

Low temperatures can prevent fusion at the junction, causing the streams to solidify before fusion, creating a cold shut.Cold shut is usually a result of a lack of fluidity of the molten metal, or a poor design of the gating system.

The best way to prevent cold shuts is to increase fluidity of the molten metal. This can be done in a few ways:

- Optimizegatingsystem to minimize narrow cross-paths and ensure short flow paths
- Increasethepouringtemperature to prevent premature solidification
- Improvegaspermeabilityof the mold (through coarser grain size, etc.)

16. Misruns are closely related to cold shuts

Misruns occur when the liquid metal is too cold to flow to the extremities of the mold cavity before freezing and solidifying.

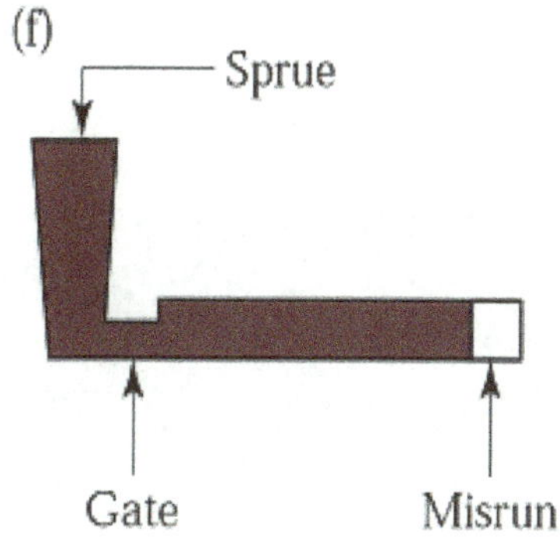

Fig.30 Misruns

The liquid metal does not completely fill the mold cavity. The misrun is the unfilled portion or space in the mold.

Causes and prevention of misruns

The reasons for premature solidification are similar to those for cold shut. If you have a misrun, check:

- Mold design
- Gating system design and
- Molten metal fluidity

17. Cold shots

Splattering during pouring of a liquid can cause solid globules to form.

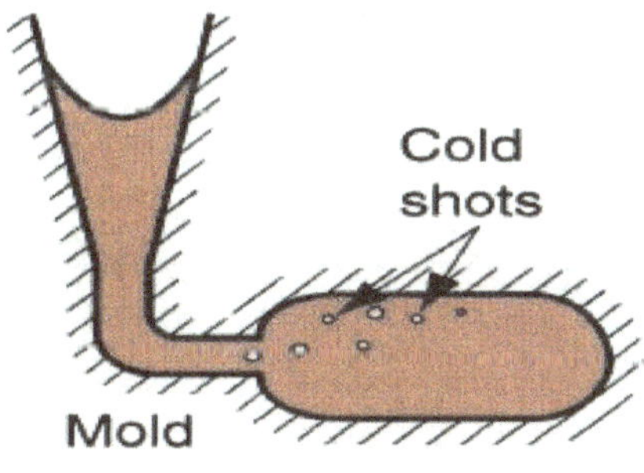

Fig.31 Cold shots

As these globules freeze, they become entrapped in the casting.

Cold shots are typically ball, drop or pearl shaped and loosely attached to the metal.

Causes and prevention of cold shots

To prevent splattering and cold shots, consider

- Modifyingpouringprocedures to minimize turbulence
- Adjustinggatingsystemdesigns to reduce gate speed

18. Slag inclusion (scab)

These irregular metallic crusts are found on the casting surface. Scabs are typically only a few millimeters thick but can be seen by the naked eye. They usually have sharp edges, irregular shapes and are firmly bonded to the casting.

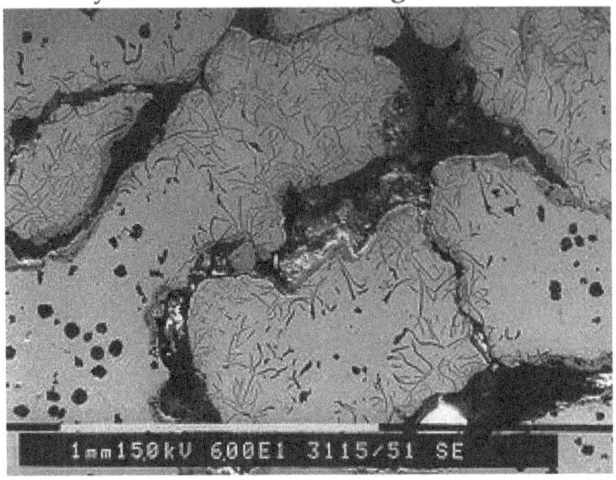

Fig.32 Scab

Scabs are closely related to rat tails and they usually appear together. Removal of scabs will typically reveal a rat tail underneath.

Causes and prevention of slag inclusion

Slag inclusion is caused when molten metal containing slag particles is poured into the mold cavities and solidifies.

Preventing slag inclusion is a simple fix. Remove slag particles from the molten metal before pouring it into the mold cavity.

You can remove slag by:

- Melting the metal with a flux, in a vacuum or in an inert atmosphere
- Adding ingredients to the mixture to cause slag to float to the top where you can easily see and remove it before pouring. Or use a special ladle that pours metal from the bottom.
- Adding a ceramic filter into the gating system

Casting shape defects and causes

These types of casting defects are related to the overall shape of the final casting.

19. Shift/mismatch

Mold shift is due to misalignment of upper (cope) and lower (drag) part of the mold. Mold shift is usually reflected as a horizontal displacement.

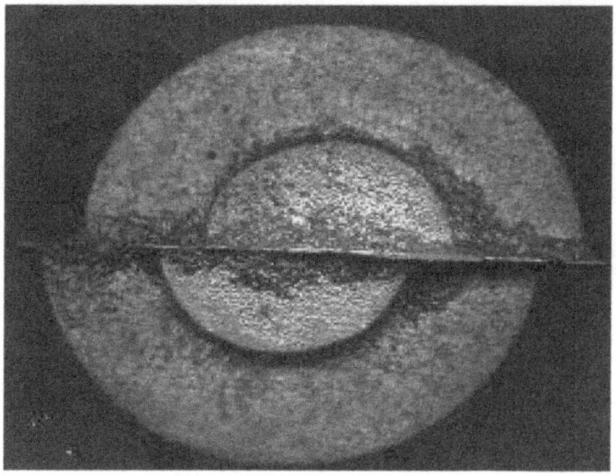

Fig.33Mold shift

Core shift is similar to mold shift, but it's the core that is misaligned, not the mold. Core shift is usually reflected as a vertical displacement.

Causes and prevention of shift

Some causes of shift can include:

- Loose box pins
- Inaccurate pattern dowel pins or
- Carelessness in placing the cope on the drag, causing misalignment

If you're experiencing shift, try checking the match plate pattern mounting and alignment. Make sure to use proper molding box and closing pins.

20. Flash, fin and burrs

Flash is one of the most frequently occurring casting defects and also a common injection molding defect.

Flash, also known as casting fin or burrs, is any unwanted and excess material attached to a cast. It's typically a thin sheet of metal that forms at the parting faces. Flash is a waste material that turns into dross after being re-melted.

Causes and prevention of flash, fin and burrs

Flash on the casting surface is due to a crack or gap on the core surface. Insufficient weight on the mold or improper clamping of the flask can lead to a gapRemedy this issue by reassembling the mold and cores. There should be enough weight on the top part of the mold so that the two parts fit together tightly.

Fig.34Flash, fin and burrs

Flash can vary from minor to very serious. If it's not too serious, manufacturers can remove flash by breaking it off with a hammer or pliers and filing it down to the parting line. However, this can be a costly process.

21. Warping

Warping is an unwanted casting deformity that can occur over time, which results in a change in the dimensions of the final product. It can happen during or after solidification.

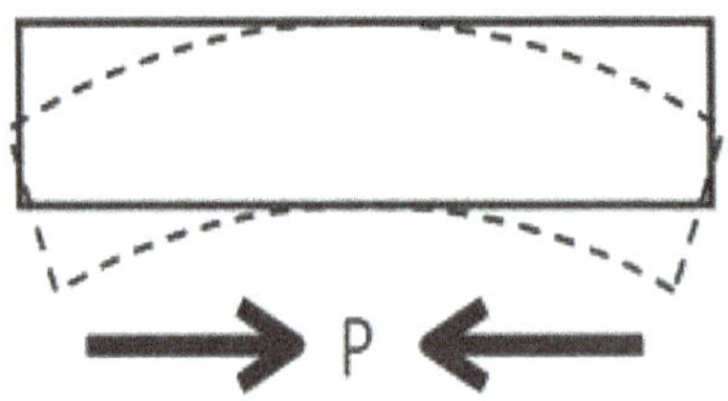

Fig.35 Warping

Causes and prevention of warping

Warping is typically a result of different rates of solidifications of different sections, which causes stress in adjoining walls. Large and flat sections are more prone to warping.

Normalizing heat treatment can remove residual stress in iron casting. A straightening between quench and aging processes might also be required for aluminum casting.

Casting manufacturing is a process in which liquefied material, such as molten metal, is poured into the cavity of a specially designed mold and allowed to harden. After solidification, the workpiece is removed from the die to undergo various finishing treatments or for use as a final product. Casting methods are typically used to create intricate solid shapes, and cast products are found in a wide range of applications, including automotive components, aerospace parts, electronics, mechanical devices, and construction supplies.

12. DIFFERENT TYPES OF CASTING AND THE CASTING PROCESS

Although casting is one of the oldest known manufacturing techniques, modern advances in casting technology have led to a broad array of specialized casting methods. Hot forming processes, such as die casting, investment casting, plaster casting, and sand casting, each provide their own unique fabrication benefits. Comparing both the advantages and disadvantages of the common types of casting processes can help in selecting the method best-suited for a given production run.

Sand Casting

Sand casting typically relies on silica-based materials, such as synthetic or naturally-bonded sand. Casting sand generally consists of finely ground, spherical grains that can be tightly packed together into a smooth molding surface. The casting is designed to reduce the potential for tearing, cracking, or other flaws by allowing a moderate degree of flexibility and shrinkage during the cooling phase of the process. The sand can also be strengthened with the addition of clay, which helps the particles bond more closely. Many automotive products, such as engine blocks and housings, are manufactured through sand casting.

Advantages of Sand Casting

Sand casting's main advantages as a casting process include:

- Relatively inexpensive production costs, especially in low-volume runs.
- The ability to fabricate large components.
- A capacity for casting both ferrous and non-ferrous materials.
- A low cost for post-casting tooling.

Disadvantages of Sand Casting

Despite its benefits, sand casting yields a lower degree of accuracy than do alternate methods and it can be difficult to sand cast components with predetermined size and weight specifications. Furthermore, this process has a tendency to yield products with a comparatively rough surface finish.

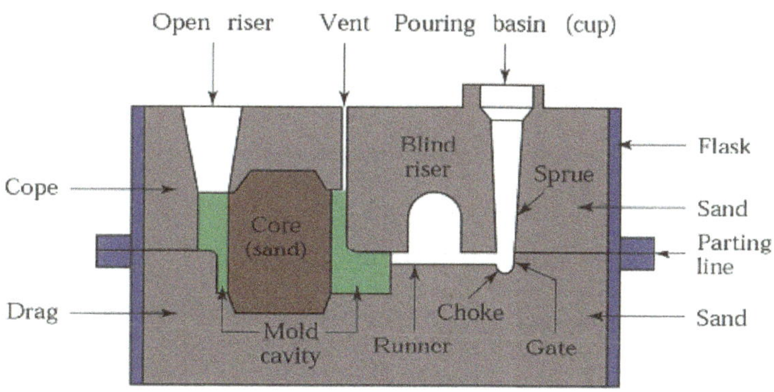

Fig.36 Sand Casting

Centrifugal casting is a process that delivers components of high material soundness. As a result, it is the technology of choice for applications like jet engine compressor cases, hydro wear rings, many military products, and other high-reliability applications. It has also proven to be a cost-effective means of providing complex shapes with reduced machining requirements and lower manufacturing costs as compared to forgings and fabrications.

Working Principle of Centrifugal Casting

In this process a molten metal is poured into the spinning mold preheated to a certain temperature. The mold is placed vertically or horizontally based on the required shape of product. Once poured it is then continued to rotate about its central axis. Due to rotational motion of the mold; a centrifugal force is acted upon the molten metal just poured into the spinning mold. This force displaces the molten metal's towards the periphery forcing them to deposit on the walls.

The molten metal is spread uniformly on to the walls of the die; thanks to the centrifugal force 100 times greater than of gravity. As the process continue with more and more metal poured into the mold; the relatively denser element tends to deposit on towards the wall while lighter element and slug deposit at the center. The mold is then left to rotate till the whole mold solidify and then other light elements like slag is separated from the center.

The whole process itself leads to reduction in defects due t slags, irregular grain structure and trapped air. The final products have closed grain structure with improved elongation, tensile strength and yield strength.

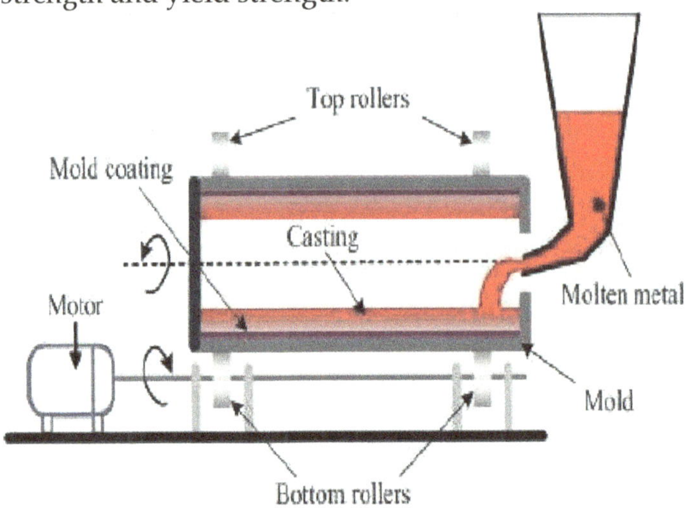

Fig.37 Centrifugal Casting

Advantages of Using Centrifugal Casting

The whole process of centrifugal casting does not rely on gates and risers. This results in continues availability of molten metal at the center during the solidification process.

Unlike conventional casting where the mold solidifies from both inside and outside; molten metal at the center ensures unidirectional solidification (Outside to Inside).

This helps get rid of defects such as blow holes, shrinkage cavity and gas pockets. The process allows for subsequent savings on initial capital (Non machinery cost) and manufacturing costs. This allows for economic advantage which gave enough flexibility to produce various shape and size of products.

The combination of deciding factors such as unidirectional solidification, solidification under pressure and impurities displacement to the central axis; results in superior quality products with high soundness compared to other industrial process. Such Advantages results in increased life and endurance of the product without fracturing.

Other benefit of centrifugal casting includes:

1. Impurities collected at the center are relatively easier to remove than other industrial or manufacturing process.
2. Mass production of symmetrical product was made possible at much lower cost.
3. It provide dense metal with mechanical soundness.
4. Relatively less temp of molten metal is required for the process saving money and energy.
5. Gates and risers are no longer needed.
6. After casting process are heavily reduced.

Types of Centrifugal Casting

The process of using centrifugal force for casting can be divided into three major parts; True centrifugal, semi-centrifugal and centrifuging.

1) True Centrifugal Casting

True centrifugal casting or commonly known as normal centrifugal casting is used to produce symmetrical hollow structure with round holes. The key feature of this process is to produce symmetrical hollow structure without using any cores. It is achieved by pure centrifugal force by rotating mold about its vertical or horizontal axis.

The shape of the mold can be circular, square, rectangular or hexagonal; as long as they are symmetrical about its vertical or horizontal axis of rotation. Centrifugal force acting on the molten metal introduced in the system; force it towards the wall of the mold / die. casting of long parts such as pipes and liner are done along the horizontal axis while for others along the vertical axis.

To avoid molten metal to take parabolic path along the mold while hardening due to gravity; the mold is subjected to high speed rotation to produce centrifugal force 100 times stronger than of gravity. The mold is then left to rotate on its axis till it solidifies unidirectionally. This process is generally used for producing large and medium sized parts such as cylinder liner, hollow pipes and bushes.

2) Semi-Centrifugal Casting

A core is placed inside the mold in a semi centrifugal casting to produce hollow structure. The process itself includes the rotation of mold along its vertical axis with the core inserted at the center. The centrifugal force from the rotation is then used to fill the mold perfectly. Here the hot metal is poured along the axis to first fill along the walls with the centrifugal force and towards center due to gravity.

A core is inserted whenever a hollow structure is needed to compose compensating the gravitational forces (dominant) at the center. This process is used to produce large Axis symmetrical products such as flywheel and gear blanks.

3) Centrifuging

This casting process is used to cast metal under high pressure for relatively small mold / die / final product or asymmetrical products in group. The process is done in group to obtain overall symmetry of the casting. Hot metal is poured into the die along the central axis through central spur using centrifugal force through radial in gates. It is an easy method for obtaining unidirectional solidification at economical costs.

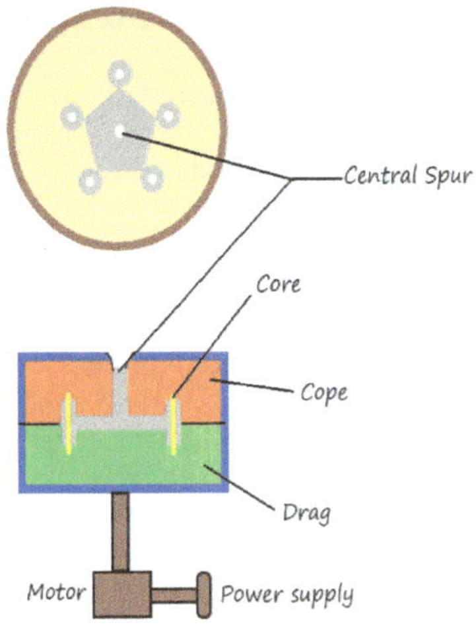

Fig.38 Centrifuging

Limitations of Centrifugal Casting

Although the process is beneficial for producing many symmetrical and asymmetrical shapes; it has some limitations to its application. Such Limitation or disadvantages of centrifugal casting are:

1. More skilled labors are required for the implementation of the process.
2. Only few defined shapes can be produced using this process.
3. Not all metal and alloys are compatible with this process.
4. High initial investment is required in machinery and tools.
5. Diameter of the inner surface is incorrect in this casting.

Applications

Different material set such as copper, nickel and all forms of steel can be used for this process. In shipbuilding centrifugal casting can be used for the manufacture of liner, vanes, parts of pump, parts of rudder and propeller, bearing housing, butterfly valves and valves body. Its other application includes piping, jet avionics and military.

Die casting is a metal casting process that is characterized by forcing moltenmetal under high pressure into a mold cavity. The mold cavity is created using two hardened tool steeldies which have been machined into shape and work similarly to an injection mold during the process. Most die castings are made from non-ferrous metals, specifically zinc, copper, aluminium, magnesium, lead, pewter, and tin-based alloys. Depending on the type of metal being cast, a hot- or cold-chamber machine is used.

Hot-chamber die casting

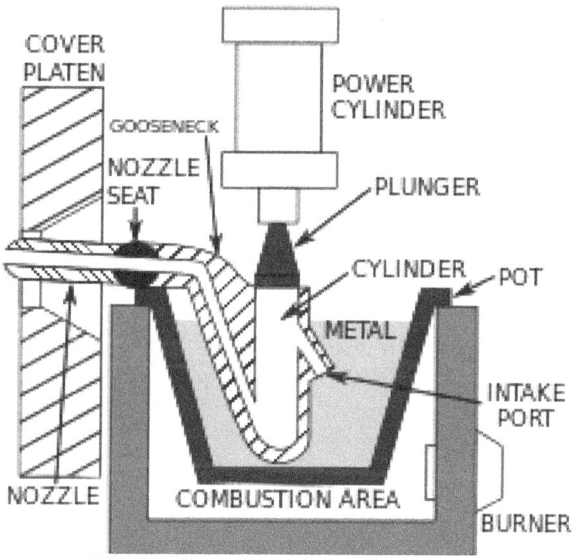

Fig.39 Hot-chamber machine

Hot-chamber die casting, also known gooseneck machines, rely upon a pool of molten metal to feed the die. At the beginning of the cycle the piston of the machine is retracted, which allows the molten metal to fill the "gooseneck". The pneumatic- or hydraulic-powered piston then forces this metal out of the gooseneck into the die. The advantages of this system include fast cycle times (approximately 15 cycles a minute) and the convenience of melting the metal in the casting machine. The disadvantages of this system are that it is limited to use with low-melting point metals and that aluminium cannot be used because it picks up some of the iron while in the molten pool. Therefore, hot-chamber machines are primarily used with zinc-, tin-, and lead-based alloys.[13]

Cold-chamber die casting

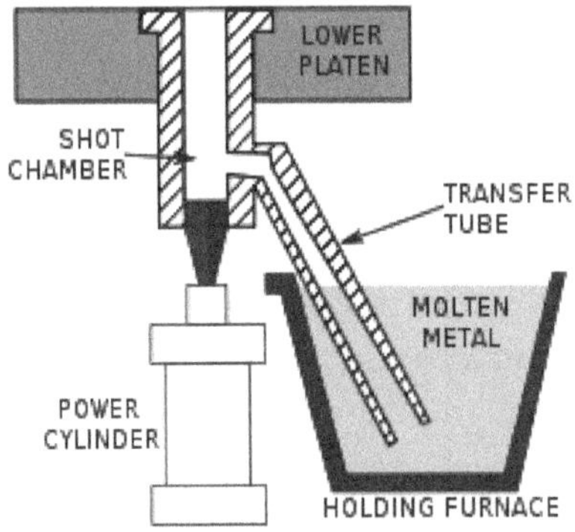

Fig.40 Cold-chamber die casting machine.

These are used when the casting alloy cannot be used in hot-chamber machines; these include aluminium, zinc alloys with a large composition of aluminium, magnesium and copper. The process for these machines start with melting the metal in a separate furnace.[14] Then a precise amount of molten metal is transported to the cold-chamber machine where it is fed into an unheated shot chamber (or injection cylinder). This shot is then driven into the die by a hydraulic or mechanical piston. The biggest disadvantage of this system is the slower cycle time due to the need to transfer the molten metal from the furnace to the cold-chamber machine.[15]

Advantages and disadvantages of die casting

Advantages of die casting:

- High productivity.
- Good dimensional accuracy.
- Good surface finish: 2-100 µinch (0.5-2.5 µm) R_a.
- Thin wall parts may be cast.
- Very economical process at high volume production.
- Fine Grain structure and good mechanical properties are achieved.
- Intricate shapes may be cast.
- Small size parts may be produced.

Disadvantages of die casting:

- Not applicable for high melting point metals and alloys (eg. steels)
- Large parts cannot be cast.
- High die cost.
- Too long lead time.
- Some gases may be entrapped in form of porosity

Die Casting Applications:

Die casting is most suitable for casting medium sized parts with complex details. Die-casting is the largest casting technique that is used to manufacture consumer, commercial and industrial products like automobiles, toys, parts of sink faucet, connector housing, gears, etc. Most die castings are done from non-ferrous metals like aluminum, magnesium, etc.

Investment Casting

In investment casting, a wax or suitable polymer pattern is coated by dipping into the refractory material slurry. Once the refractory material coating is hardened then this dipping process is repeated several times to increase the coating thickness and its strength. Once the final coating is hardened the wax is melted out and molten metal is poured into the cavity created by the wax pattern. Once the metal solidifies within the mould, metal casting is removed by breaking the refractory mould.

Investment Casting Process Steps

The following steps are what we follow to produce parts with complex geometries and intricate details.

1. **Create wax pattern**: A pattern that replicates the finished part geometry is made using one of two primary methods:
   a. Build a wax injection die
   b. 3D print pattern
      (i) If a wax injection die will be used, the first step involves designing and building a metal die from Aluminum. This die creates a wax replica of the desired part by injecting melted wax under high pressure into the cavity. The die can be made as a simple one cavity manual tool or a complex multi-cavity automatic tool depending on volume requirements.
      (ii) If a 3D printed pattern will be used, a CAD model which contains the part geometry is sent to a printer and the part is printed.
2. **Wax Assembly**: Next, the wax patterns are assembled onto runners and into a finished tree which is ready to be dipped.
3. **Slurry Coating**: The assembly is then dipped into a high-grade ceramic slurry to build a ceramic shell around the wax tree.
4. **Stuccoing**: After the slurry coating is done, particles of sand are dropped onto the surface of the wet tree assembly. This helps to thicken and strengthen the layer of coating on the wax assembly surface.
5. **Repeat Steps 5 and 6**: Repeat steps 5 and 6 until the assembly achieves the desired shell thickness. The assembly is then allowed to set and dry.
6. **Dewaxing**: The wax inside the newly built shell is now removed. Dewaxing is done using a steam-dewaxing autoclave or flash fire furnace.
7. **Casting:** Now the desired molten metal is poured into the pre-heated mold cavity.
8. **Cooling**: The mold then sits to allow the molten metal to cool and solidify which then becomes the final casting.

9. **Shell Removal**: The shell material is then removed through processes hammer knockout, vibration, and steel grit blasting.
10. **Cut Off**: The finished parts are then cut free from the gating and runner system.
11. **Finishing**: Various finishing techniques are then employed including grinding, sand blasting and coating to achieve the final surface needed.
12. **Testing**: Once the finishing operations are done, the parts are inspected for surface and sub-surface defects. Visual and fluorescent penetrant inspection is done for surfaces and X-ray is employed for sub-surface defect identification.

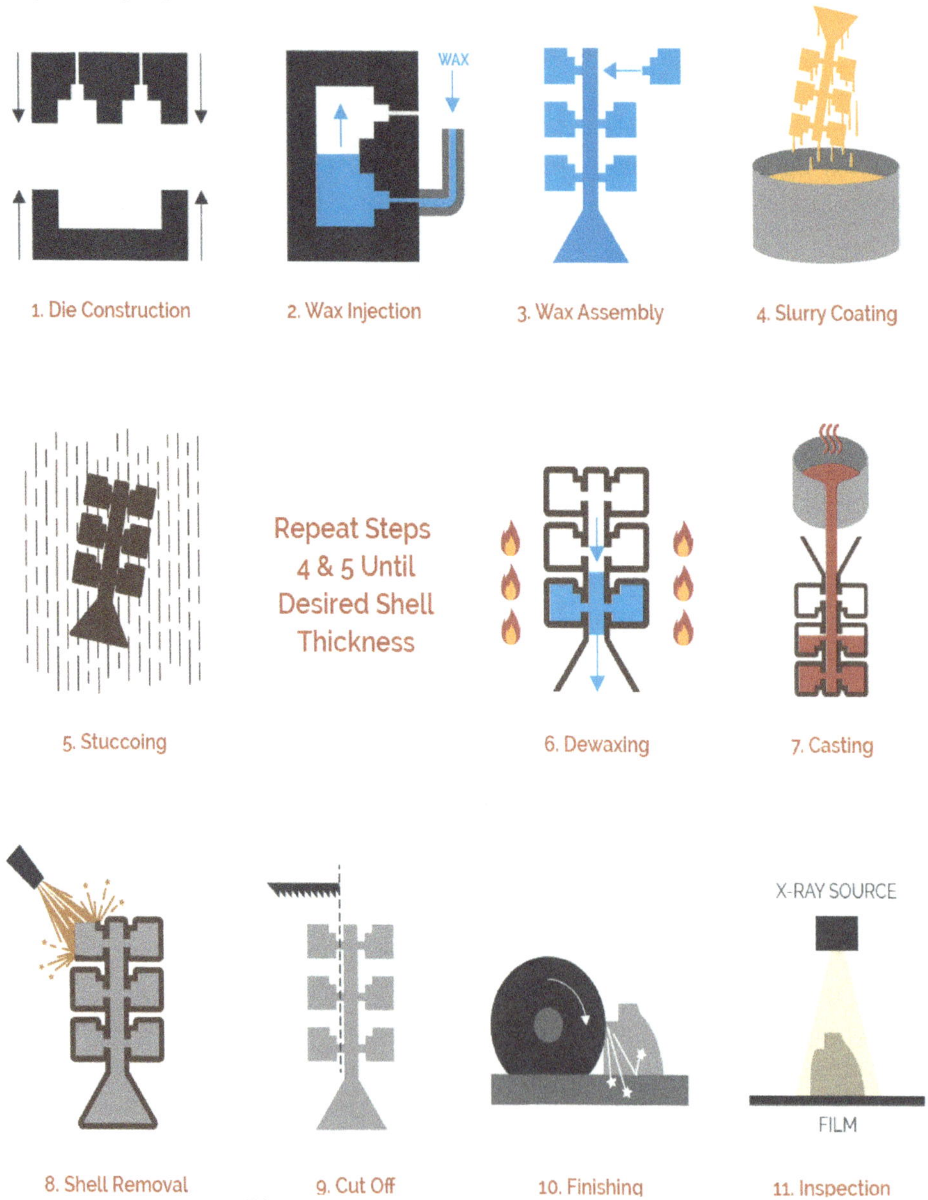

Fig.41Investment Casting Process

Advantages of Investment casting

- Parts with extremely complex shapes and intricate features can be cast as a single piece using investment casting
- With short length or shallow depth feature, thin sections down to 0.40 mm (0.015 in) can be cast without cold shut defects
- Lost wax casting has excellent dimensional accuracy and tighter tolerances of 0.075 mm (0.003 in) are easily achievable.
- Compared to similar manufacturing processes, Investment casting can achieve excellent surface finish without any post-processing. Typically around 1.3 - 0.4 microns RMS Ra
- It offers almost unlimited freedom in terms of investment cast materials, but most common materials used include Aluminium alloy, cast iron and non-ferrous alloys. The process is particularly attractive for high-temperature alloys.
- Draft on walls are not required but if a master die is used to make wax patterns then draft on the face would help the pattern making process easier.
- Since there are no parting lines, the cast would not have any flash. But the wax patterns might have parting lines from the master die.
- Additional machining can be eliminated or reduced and allowance of as little as 0.4 to 1 mm (0.015 to 0.040 in.) usually enough.
- Excellent dimensional precision can be achieved in combination with very smooth as-cast surfaces. These capabilities are especially attractive when making products from the high-melting temperature, difficult-to-machine metals that cannot be cast with plaster- or metal mould processes.
- The wax used can usually be recovered for reuse.

Disadvantages of Investment casting

- Compared to other methods of metal casting, investment casting involves many complex steps making the process relatively expensive. But some of the steps can be automated for certain products. It can be more expensive than die casting or sand casting, but per-unit costs decrease with large volumes.
- The high cost of dies to make patterns has traditionally limited investment casting to large production quantities
- The high cost is also due to specialized equipment requirement, costly refractory material, and high labor cost
- Parts are difficult to cast if they require cores, got holes smaller than 1.6 mm or deeper than 1.5 times the diameter

Applications

We are proud to produce complex investment cast parts with superior surface finishes in the industries fastest lead-times for the following markets:

- Aerospace
- Defense
- Medical
- Electronics
- Automotive
- Oil and Gas
- Agriculture
- Commercial

Shell molding

Shell molding, also known as shell-mould casting, is an expendable mold casting process that uses a resin covered sand to form the mold. As compared to sand casting, this process has better dimensional accuracy, a higher productivity rate, and lower labor requirements. It is used for small to medium parts that require high precision

The process of creating a shell mold consists of six steps:

1. Fine silica sand that is covered in a thin (3–6%) thermosetting phenolic resin and liquid catalyst is dumped, blown, or shot onto a hot pattern. The pattern is usually made from cast iron and is heated to 230 to 315 °C (450 to 600 °F). The sand is allowed to sit on the pattern for a few minutes to allow the sand to partially cure.
2. The pattern and sand are then inverted so the excess sand drops free of the pattern, leaving just the "shell". Depending on the time and temperature of the pattern the thickness of the shell is 10 to 20 mm (0.4 to 0.8 in).
3. The pattern and shell together are placed in an oven to finish curing the sand. The shell now has a tensile strength of 350 to 450 psi (2.4 to 3.1 MPa).
4. The hardened shell is then stripped from the pattern.
5. Two or more shells are then combined, via clamping or gluing using a thermoset adhesive, to form a mold. This finished mold can then be used immediately or stored almost indefinitely.
6. For casting the shell mold is placed inside a flask and surrounded with shot, sand, or gravel to reinforce the shell.

The machine that is used for this process is called a shell molding machine. It heats the pattern, applies the sand mixture, and bakes the shell.

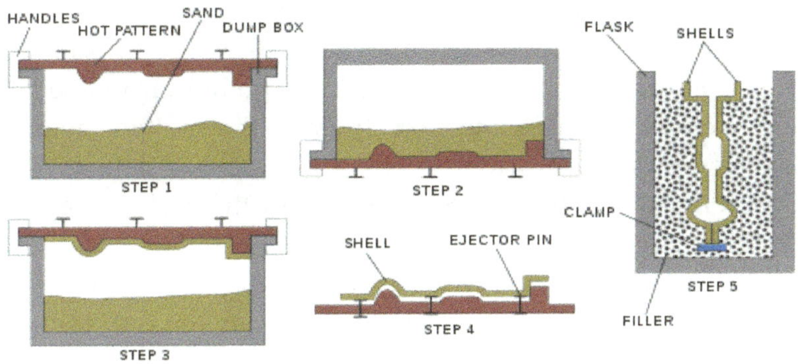

Fig.42 Shell molding Process

13. GATING SYSTEM AND TYPES

It is nothing but the basic design, which is needed to construct a smooth and proper filling of the mold cavity of the casting without any discontinuity, voids or solid inclusions. A proper method of gating system is that it leads the pure molten metal to flow through a ladle to the casting cavity, which ensures proper and smooth filling of the cavity. This depends on the layout of the gating channels too, such as the direction and the position of the runner, sprue and ingates.

Objective of the Gating System:

The four main points, which enables a proper gating system, are:

- Clean molten metal.
- Smooth filling of the casting cavity.
- Uniform filling of the casting cavity.
- Complete filling of the casting cavity.

The mold cavity must be filled with a clean metal so that it prevents the entry of slag and inclusions into the mold cavity, which in turn minimizes the surface instability. If the mold has smooth filling then it helps to reduce the bulk turbulence. If it has a uniform filling it means that the casting fill is in a controlled manner. Complete filling of the cavity makes the metal thin with minimum resistance at the end sections.

Procedure:

The main elements needed for the gating system are as follows:

- Pouring basin or bush.
- Sprue or down Sprue.
- Sprue Well
- Runner
- Ingate
- Ladle
- Slag trap or filter.

The characteristics of each element are mentioned below:

- **Pouring basin:** This is otherwise called as bush or cup. It is circular or rectangular in shape. It collects the molten metal, which is poured, from the ladle.
- **Sprue:** It is circular in cross section. It leads the molten metal from the pouring basin to the sprue well.
- **Sprue Well:** It changes the direction of flow of the molten metal to right angle and passes it to the runner.
- **Runner:** The runner takes the molten metal from sprue to the casting. Ingate: This is the final stage where the molten metal moves from the runner to the mold cavity.
- **Slag trap:** It filters the slag when the molten metal moves from the runner and ingate. It is also placed in the runner.
- **Types of Gating Systems:**
- The gating system also depends on the direction of the parting plane, which contains the sprue, runner and the ingate. They are as follows:
- **Horizontal Gating System:** This is used most widely. This type is normally applied in ferrous metal's sand casting and gravity die-casting of non-ferrous metals. They are used for flat casting, which are filled under gravity.
- **Vertical Gating System:**This is applied in tall castings were high-pressure sand mold, shell mold and die-casting processes are done. Top Gating System: this is applied in places where the hot metal is poured form the top of the casting. It helps directional solidification of the casting from top to bottom. It suits only flat castings to limit the damage of the metal during the initial filling.
- **Bottom Gating System:** it is used in tall castings where the molten metal enters the casting through the bottom.
- **Middle Gating System:** It has the characteristics of both the top and bottom.

Gating Design and Analysis

A well-designed gating system is a perquisite for achieving perfect casting. The key objective of a gating system is to ensure smooth and complete flow of molten metal from ladle to the casting cavity.

Elements of gating system

The main components include pouring basin, Sprue, well, runner and ingate. Gating systems can be classified based on the position of the parting plane and also on the basis of ingate position as follows,

i) **Horizontal systems-** Suitable for flat castings filled under gravity and commonly used in sand casting of ferrous metals and gravity die casting of non-ferrous metals.

ii) **Vertical systems-** Suitable for tall castings and used in high pressure sand and shell mold and die casting processes.

iii) **Top gating systems-** wherein the molten metal enters at the top of the casting promoting directional solidification from bottom to top of the casting.

iv) **Bottom gating systems-** wherein the molten metal enters from the bottom and fills the mold slowly with minimal

v) **Middle gating systems-** these combine the features of top and bottom systems.

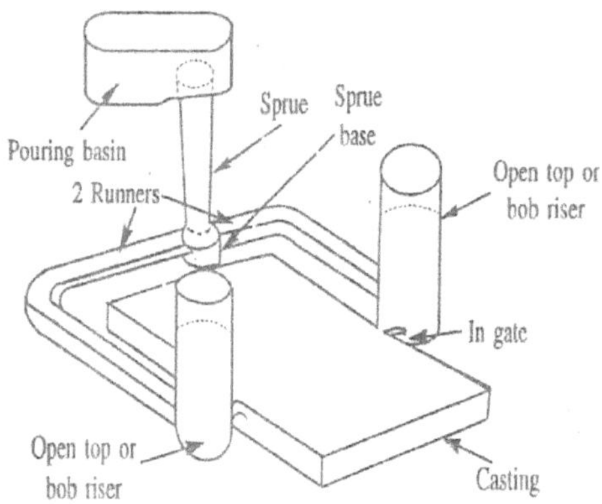

Fig.43Elements of gating system

The number and position of ingate(s) is very important in designing the gating systems. The diagram given below describes the different considerations in the channel layout in horizontal gating system with side ingates.

Heuristics for ingate location

i) **Side feeders-** When side feeders are used their efficiency can be improved by filling the first stream of hot molten metal through ingates, thereby reducing the fettling effort and subsequent marks on the casting.

ii) **Thick sections -** This allows molten metal to flow to other sections with minimal cooling reducing breakage during fettling of ingates.

iii) **Clear path -** In case of sand casting it is essential to allow the metal to flow with minimal obstructions or direction changes to curb turbulence. In such systems ingates should never be placed opposite to a core.

iv) **Low free fall -** here the ingates are located where the free fall of molten metal in the mold cavity is low thereby minimizing oxidation and erosion at the point of impact.

It is also essential to use adequate number of ingates ensuring that the distance of flow between the ingate and any point filled by that ingate is always less than the fluidity distance. Usually the sprue directs the flow of metal from the basin to the runners and the ingates. The location of sprues depends on the following factors,

i) **Flow distance -** It must minimize the total flow distance with the gating channel to give maximum yield and minimum heat loss.

ii) **Heat concentration -** It must be located away from hot spots in the casting

iii) **Mold layout -** It should be positioned to minimize the size of the bounding box, which encloses the entire casting.

iv) **Hick sections-** This allows molten metal to flow to other sections with minimal cooling reducing breakage during fettling of ingates.

Optimal Filling Time

Filling time is very crucial in casting as it decides the final quality of product. A slow fill gives cold shuts and misruns while a fast fill can lead to solid and gaseous inclusions. The ideal filling time is a function of cast metal, weight, minimum section thickness and pouring temperature and can be expressed as, tf = K0 (Kf Lf / 1000) (Ks + Kt t / 20) (Kw W)P

Where tf is the casting time, W is the weight in kg, t is the section thickness in mm and Lf is the fluidity length in mm, whereas K0, Kf , Ks , Kt , Kw are all coefficients. The value of these coefficients can be set to achieve a perfect metal-process combination in specific castings.

The velocity of molten metal also crucial role in deciding the optimal filling time. The velocity usually varies within the gating channels and the mold cavity. It depends majorly on two factors,

- Metallostatic pressure
- Gating ratio

Gating Element Design

A gating system is designed to fill the casting in prescribed time keeping a constant level of liquid metal in the basin to achieve a controlled flow rate via the choke, a small cross section in system that regulates the flow rate.

The dimensions and shape of different elements in the gating systems is determined as follows,

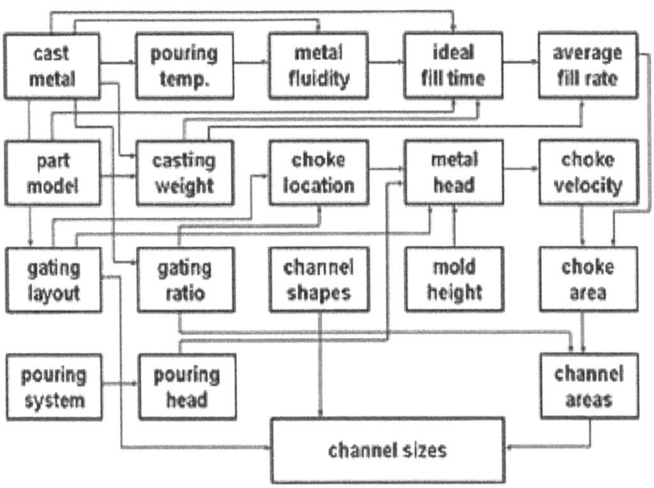

Fig.44Gating Element Design

- **i) Sprue -** It is a circular cross-section minimizing turbulence and heat loss and its area is quantified from choke area and gating ratio. Ideally it should be large at top and small at bottom.

- **ii) Sprue well:** It is designed to restrict the free fall of molten metal by directing it in a right angle towards the runner. It aids in reducing turbulence and air aspiration. Ideally it should be shaped cylindrically having diameter twice as that of sprue exit and depth twice of runner.
- **iii) Runner -** Mainly slows down the molten metal that speeds during the free fall from sprue to the ingate. The cross section are of a runner should be greater than the sprue exit. It should also be able to fill completely before allowing the metal to enter the ingates. In systems where more than one ingate is present, it is recommended that the runner cross section area must be lowered after each ingate connection to ensure smooth flow.
- **iv) Ingate:** It directs the molten metal from the gating system to the mold cavity. It is recommended that ingate should be designed to reduce the metal velocity; they must be easy to fettle, must not lead to a hot spot and the flow of molten metal from the ingate should be proportional to the volume of casting region.

Assessment of Gating Design

Different combinations of gating system designs and mold filling are carried out to completely eliminate filling defects. A gating system can be assessed using the following criteria,

- **i) Mold filling time -**The actual mold filling time must be close to the optimal filling time of the gating system. This can be expressed as, CG1 = 1 - (| tf-actual - tf-optimal |) / tf-optimal
- **ii) Ingatevelocity:** The velocity of molten metal emerging from the ingate must be as low as possible to minimize turbulence. It can be expressed as, CG2 = 1 - (Vingate / Vcritical)
- **iii) Impingement -**Mold erosion is caused by the velocity and direction of the first stream of molten metal emerging from the ingate and striking the mold face. A fast stream should be avoided and can be expressed as, CG3 = Vimp-limit / (Vimp (nimp. nf))
- **iv) Gating yield -**The volume of the gating system must be minimized to increase the yield. The criterion is given by: CG4 = Ncvc / (Ncvc + vg)
- **v) Fettling -** The ingate size must be small in comparison to the connected part of casting to avoid breakage and cracks.

Validation of gating design

The gating design can be validated using the following techniques,

i) **Shop floor trials -** Samples are produced using the actual materials and process that would be finally employed. These are observed for surface, sub-surface and internal quality and destructive and non-destructive testing methods.

ii) **High-speed radiography -** The mold filling phenomenon is recorded using a high speed x-ray camera and observed of any defects.

iii) **Open mold -** In this a portion of cope above casting cavity is cut and a flow of molten metal stream form ingate and gradual filling of the mold is recorded using a video camera.

iv) **Contact wire sensing -** In this contact wires are placed in different parts of the mold and when the metal touches a particular wire, the metal flow gets recorded by a multichannel recorder.

v) **Water in transparent mold -** Flow of water in a transparent mold is a very useful tool. Addition of color marker, oil droplets can help in better visualization and recording velocities in different sections.

Runners /Sprue

Runners are connected channels that convey the molten metal to different parts of the mould.

A Sprue is a hole in the mold where the molten metal enters. On the outside it is funnel shaped and on the inside it connects to the gate or gates, sometimes called a gating system. There can be a deeper part at this junction to trap any loose sand before the metal enters the gating system.

Risers

A riser may look like a Sprue, but it has a different function. Instead of metal being poured in, it rises from the mold cavity as the Sprue is filled. It tells you when to stop pouring as well as provides molten metal to thick parts of the mold to prevent shrink cavities as it cools.

Top riser: They are open to the atmosphere.ϖIt is the most conventional and convenient to make.ϖThe top being open loses heat to the atmosphere by radiation and convection.

Blind riser: They are completely concealed inside the mould cavity.ϖIt is surrounded by the mould sand. Therefore, it loose heat slowly and effectively.ϖIt can be located more conveniently than the open riser.

Internal riser: They are enclosed on all sides by the casting such that heat from the casting keeps the metal in the riser hot for a long time. These are normally used for castings, which are cylindrical shaped or have hollow cylindrical portions.

UNIT - II

WELDING PROCESS

1. INTRODUCTION

Welding is a process of joining two metal pieces as a result of significant diffusion of the atoms of the welded pieces into the joint (weld) region. Welding is carried out by heating the joined pieces to melting point and fusing them together (with or without filler material) or by applying pressure to the pieces in cold or heated state.

Advantages of welding:

- Strong and tight joining
- Cost effectiveness
- Simplicity of welded structures design
- Welding processes may be mechanized and automated

Disadvantages of welding:

- Internal stresses, distortions and changes of micro-structure in the weld region
- Harmful effects: light, ultra violate radiation, fumes, high temperature.

Applications of welding:

- Buildings and bridges structures
- Automotive, ship and aircraft constructions
- Pipe lines
- Tanks and vessels
- Railroads
- Machinery elements.

Welding is broadly classified into two types:

I. Plastic Welding:

In this case the two pieces of metals to be joined are heated to plastic state and welding is completed by forcing them. In this case no filler material is added. In this process, pressure is applied. So it is also called as pressure welding.

Forge Welding:

It is the oldest method of joining two pieces of metals. In this case heat is created by blacksmith's fire and the two pieces of metals to be joined are heated and brought to plastic state. Then they are superimposed and hammered together to form the joint.

Resistance Welding:

In this case a heavy electric current is passed through the pieces of metals to be joined. Then because of electric resistance, the metals are heated to plastic state and by applying force, welding can be completed. In this case, no additional filler material is required and the metal pieces are pressed in the two copper electrodes.

II. Fusion Welding:

In this case the two pieces of metal are heated and brought to molten state and filler material is added and allowed to solidify. In this process, no pressure is applied. So it is also called as non pressure welding.

Welding processes

- Arc welding
  - Carbon Arc Welding
  - Shielded Metal Arc Welding (SMAW)
  - Submerged Arc Welding (SAW)
  - Metal Inert Gas Welding (MIG, GMAW)
  - Tungsten Inert Gas Arc Welding (TIG, GTAW)
  - Electroslag Welding (ESW)
  - Plasma Arc Welding (PAW)
- Resistance Welding (RW);
  - Spot Welding (RSW)
  - Flash Welding (FW)
  - Resistance Butt Welding (UW)
  - Seam Welding (RSEW)
- Gas Welding (GW);
  - Oxyacetylene Welding (OAW)
  - Oxyhydrogen Welding (OHW)
  - Pressure Gas Welding (PGW)
- Solid State Welding (SSW)
  - Forge Welding (FOW)
  - Cold Welding (CW)
  - Friction Welding (FRW)
  - Explosive Welding (EXW)
  - Diffusion Welding (DFW)
  - Ultrasonic Welding (USW)
- Thermit Welding (TW)
- Electron Beam Welding (EBW)
- Laser Welding (LW)

Different Types of Welding Joints:

Tee

Tee welding joints are formed when two members intersect at a 90° angle which makes the edges come together in the center of a plate or component. Tee Joints are considered a type of fillet weld, and can also be made when a pipe or tube is welded onto a base plate. Extra care is required to ensure effective penetration into the roof of the weld.

Lap

Lap welding joints are used most often to joint two pieces with differing thicknesses together. Also considered a fillet type, the weld can be made on one or both sides. A Lap Joint is formed when 2 pieces are placed in an over lapping pattern on top of each other.

Edge

Edge welding Joints are often applied to sheet metal parts that have flanging edges or are placed at a location where a weld must be made to attach to adjacent pieces. Being a groove type weld, Edge Joints, the pieces are set side by side and welded on the same edge. For heavier applications filler metal is added to melt or fuse the edge completely and to reinforce the plate.

Corner

Being one of the most popular welds in the sheet metal industry the Corner welding joint is used on the outer edge of the piece. This weld is a type of joint that comes together at right angles between two metal parts to form an L. These are common in the construction of boxes, box frames and similar fabrications.

Butt

Being the universally accepted method for attaching a pipe to itself it's also used for valves, flanges, fittings, and other equipment. A butt welding joint is also known as a square grove weld. It's the easiest and probably the most common weld there is. It consists of two flat pieces that are side by side parallel. It's a very affordable option.

Welding joints can be prepared in numerous ways including:

- Casting
- Shearing
- Machining
- Forging
- Filing
- Stamping
- Oxyacetylene cutting (thermal cutting process)
- Routing
- Grinding
- Plasma arc cutting (thermal cutting process)

2. GAS WELDING

Gas welding is a fusion welding process in which the heat for welding is obtained by the combustion of oxygen and fuel gas. The fuel gas may be acetylene, hydrogen, propane or butane.

An intense gas flame is thus produced which melts the edges of the parts to be welded. The molten metal is allowed to flow to Solidify together and continuous joint is obtained. The gas welding is particularly suitable for joining metal sheets and plates having thickness of 2 to 50 mm. An additional metal called filler material is used for thickness more than 15mm. This filler metal is used in the form of welding rod.

The Composition of filler rod is usually same as that of base metal. The filler metal is used to fill up the cavity made during edge preparation. A flux material is also used during welding to remove impurities and oxides present on the metal surfaces to be joined.

Different combinations of gases are used to produce hot gas flame, e.g., Oxygen and acetylene, oxygen and hydrogen, oxygen and propane, air and acetylene etc.

The combination of oxygen and acetylene is most widely used. This combination burns to produce a highest flame temperature of about 3200°C. Such a flame produced is known as Oxy-acetylene flame.

Oxy-acetylene flame is used for welding metals, having high melting temperature such as mild steel, high carbon steel etc. On the other hand, Oxy-hydrogen flame is used for welding metals having low melting temperature such as aluminum, lead, magnesium, etc.

Parts of a gas welding system

Gas welding requires a system where two gases, oxygen, and fuel gas, can be safely mixed and burnt together. It also requires a mechanism for the welder to control the flames.

To satisfy all these conditions, gas welding equipment consists of:

A fuel cylinder: The fuel cylinder holds the fuel required for the combustion. The cylinder is sealed and is made from heavy-gauge steel to prevent the compressed fuel from weakening the cylinder.

An oxygen cylinder: The oxygen cylinder holds the compressed oxygen required for the weld. Both the oxygen and fuel cylinders are made to withstand the pressure from the respective gases.

Control valves: Both gases have their separate control valves. A control valve is used to control the amount of gas being released from the cylinder. Control valves are also crucial for controlling the fuel to oxygen ratio.

Mixer chamber: It is a housing in which the fuel and the oxygen mix. The control valves are used to regulate the flow of gases from the cylinder to the mixer chamber.

Welding torch: A welding torch contains the mixer chamber and the control values. At the other end of the torch is a nozzle where the fuel-oxygen mixture is burnt together.

Advantages of gas welding

- Certain key features of gas welding make it so popular in the field. We will discuss some of them here.
- Weld ferrous and non-ferrous metals: One of the great strengths of gas welding is that it can weld non-ferrous and ferrous metals together.
- No requirement for electricity: When we compare gas welding to the other popular weld methods like arc welding, gas welding does not require any electricity to operate. Hence, you can use gas welding in places that do not have access to electricity.
- Cheap Equipment costs: The initial capital for gas welding is very low as it doesn't require specialized machinery.
- Doesn't require specialized labor: Gas welding doesn't require highly specialized labor. This makes it easier to find gas welders, as well as keeps the labor charge low.
- Portable equipment: The whole setup for Gas Welding is easily portable.

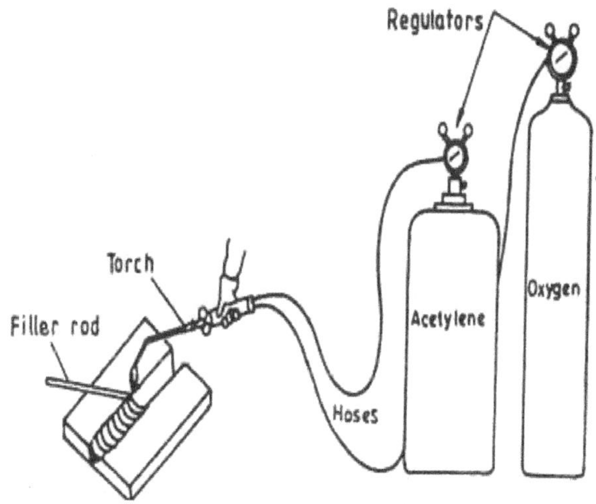

Fig.1. Gas welding

Disadvantages of gas welding

The popular term "no one-size-fits-all" is true for gas welding too. These are some of the disadvantages of using gas welding. These include:

- Not suitable for thick sections
- Cannot be used for high strength steel
- The slow rate of heating
- Cannot reach the temperatures of arc welding
- Doesn't have a dedicated flux shielding system

Application of gas welding

- Gas welding is used in a variety of industries. Here is a list of some of the most common applications of gas welding.
- Repair works: One of the most common applications of gas welding is for repair works.
- Fabrication of sheet metal: Thin to medium sheet metals are easily weld using gas welding.
- Aircraft industry: Oxy-Acetylene welding is used in joining various aircraft parts.
- Automotive industry: Used to weld parts of the frame and the chassis.
- Joining High carbon Steel: Gas welding is very effective in melting high carbon steel.

Types of Gas Welding

Oxy-Acetylene Welding

Oxy-acetylene welding uses a mixture of acetylene gas and oxygen gas to feed the welding torch. Oxy-acetylene welding is the most commonly used gas welding technique. This gas mixture also provides the highest flame temperature of available fuel gases, however acetylene is generally the most expensive of all fuel gases. Acetylene is an unstable gas and requires specific handling and storage procedures.

Oxy-Gasoline Welding

Pressurized gasoline is used as a welding fuel where fabrication costs are an issue, particularly in locations where acetylene canisters are not available. Gasoline torches can be more effective than acetylene for torch-cutting thick steel plates. The gasoline can be hand-pumped from a pressure cylinder, a common practice by jewelry makers in impoverished areas.

MAPP Gas Welding

Methylacetylene-propadiene-petroleum (MAPP) is a gas mixture that is much more inert than other gas mixtures, making it safer for hobbyists and recreational welders to use and store. MAPP can also be used at very high pressures, allowing it to be used in high-volume cutting operations.

Butane/Propane Welding

Butane and propane are similar gases that can be used alone as fuel gases or mixed together. Butane and propane have a lower flame temperature than acetylene, but are less expensive and easier to transport. Propane torches are more frequently used for soldering, bending and heating. Propane requires a different type of torch tip to be used than an injector tip because it is a heavier gas.

Hydrogen Welding

Hydrogen can be used at higher pressures than other fuel gases, making it especially useful for underwater welding processes. Some hydrogen welding equipment works off electrolysis by splitting water into hydrogen and oxygen to be used in the welding process. This type of electrolysis is often used for small torches, such as those used in jewelry making processes.

Oxy-fuel cutting

Before cutting, the cutting torch has to pre-heat the steel to ignition temperature at the starting point. At this temperature of around 960°C (depending on type of alloy), the steel has lost protective properties against oxygen and is still solid. Pure oxygen is then directed through the nozzle at the heated area. This fine and high pressure oxygen stream changes pre-heated and unprotected steel into oxidized liquid steel by an exothermic reaction.

This slag has a lower melting point than steel, so the oxygen stream can blow the liquid slag out of the cavity without affecting the non-oxidized solid steel. This exothermic reaction is a continuous process and creates a cut as the torch moves. To keep the exothermic reaction working, the cutting torch keeps the steel heated during cutting. Only metals whose oxides have a lower melting point than the base metal itself can be cut with this process. Otherwise as soon as the metal oxidizes it terminates the oxidation by forming a protective crust. Only mild steel and some low alloys meet the above conditions and can be cut effectively with the oxy-fuel process.

Oxy-fuel cutting is a time consuming process, especially cutting thick walled materials. The shortest distance through the material is a cut perpendicular to the wall. In case of a beveled cut (an angled cut for weld preparation) the cutting torch will be at an angle and the heat transfer to the material will be less than that of a perpendicular cut and the cut length will be greater. This requires automation of speed control to maintain an accurate angled cut.

Using an extra oxy-fuel torch (pre-heat torch) can be installed perpendicular to the material and running in front of the cutting torch can improve the cutting speed significantly (100% at 70 degrees cutting angle and 50% at 45 degrees cutting angle). To avoid unnecessary gas consumption this pre-heater should only be activated for angles steeper than 30°.

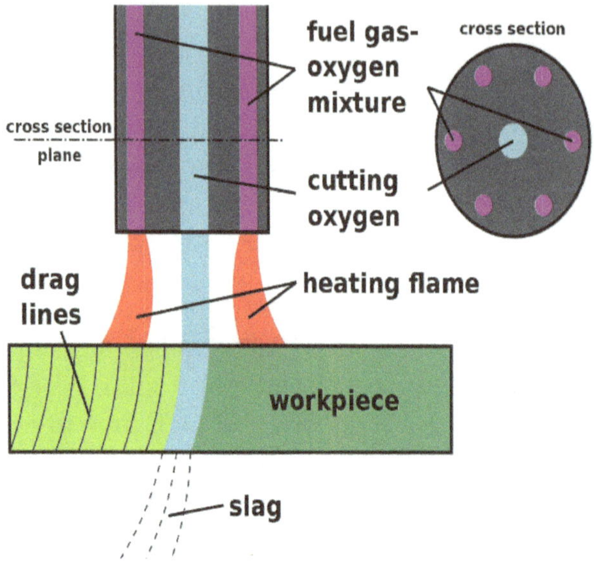

Oxy-Acetylene Welding:

When a combination of oxygen and acetylene is used in correct proportions, to produce an intense gas flame, the process is known as Oxy-acetylene welding.

An Oxy-acetylene gas flame has a temperature of about 3200°C and thus can melt all commercial available metals. A filler rod of the same material is used to fill up the cavity made during edge preparation, if metal thickness is more than 15 mm. A flux is used to remove impurities and oxides present on the metal surface.

To ignite flame, open the acetylene control valve of the welding torch. The necessary oxygen is drawn from the atmosphere to burn acetylene partially.

The oxygen control valve is then open to adjust the required volume of acetylene and oxygen mix and burn. The three different types of gas flames produced by changing the mix volumes.

Standard time and cost calculations

Labor or over head = Labor or over head cost / Deposition rate x operating factor

Electrode = Electrode cost/ Deposition rate

Gas = Cost of gas x gas flow rate / Deposition rate

Welding time = Size of weld / Deposition rate x operating factor

3. ARC WELDING

Arc welding is a fusion welding process used to join metals. An electric arc from an AC or DC power supply creates an intense heat of around 6500°F which melts the metal at the join between two work pieces.Electric arc is luminous electrical discharge between two electrodes through ionized gas.

Any arc welding method is based on an electric circuit consisting of the following parts:

- Power supply (AC or DC);
- Welding electrode;
- Work piece;
- Welding leads (electric cables) connecting the electrode and work piece to the power supply.
- Electric arc between the electrode and work piece closes the electric circuit. The arc temperature may reach 10000°F (5500°C), which is sufficient for fusion the work piece edges and joining them.
- When a long join is required the arc is moved along the joint line. The front edge of the weld pool melts the welded surfaces when the rear edge of the weld pool solidifies forming the joint.

When a filler metal is required for better bonding, filling rod (wire) is used either as outside material fed to the arc region or as consumable welding electrode, which melts and fills the weld pool. Chemical compositions of filler metal is similar to that of work piece.

Molten metal in the weld pool is chemically active and it reacts with the surrounding atmosphere. As a result weld may be contaminated by oxide and nitride inclusions deteriorating its mechanical properties. Neutral shieldinggases (argon, helium) and/or shieldingfluxes are used for protection of the weld pool from atmospheric contamination. Shields are supplied to the weld zone in form of a flux coating of the electrode or in other forms.

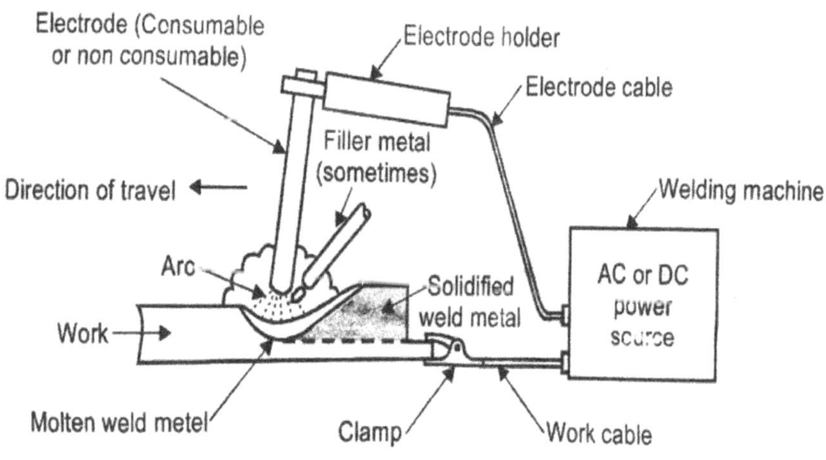

Fig.2. Arc welding

Advantages

There are a number of advantages to using arc welding compared with many other formats:

- Cost – equipment for arc welding is well-priced and affordable, and the process often requires less equipment in the first place because of the lack of gas

- Portability - these materials are very easy to transport
- Works on dirty metal
- Shielding gas isn't necessary - processes can be completed during wind or rain, and spatter isn't a major concern

Disadvantages

There are a few reasons why some people look to other options beyond arc welding for certain kinds of projects. These downsides can include:

- Lower efficiency - more waste is generally produced during arc welding than many other types, which can increase project costs in some cases
- High skill level - operators of arc welding projects need a high level of skill and training, and not all professionals have this
- Thin materials - it can be tough to use arc welding on certain thin metals

Applications of Arc Welding:

- Shipbuilding
- Automotive industries
- Construction industries
- Mechanical industries

4. CONSUMABLE ELECTRODE METHODS

Metal Inert Gas Welding (MIG) and Metal Active Gas Welding (MAG)

Also known as **Gas Metal Arc Welding (GMAW)**, uses a shielding gas to protect the base metals from contamination.

Shielded Metal Arc Welding (SMAW)

Also known as manual metal arc welding (MMA or MMAW), flux shielded arc welding or stick welding is a process where the arc is struck between the metal rod (electrode flux coated) and the work piece, both the rod and work piece surface melt to form a weld pool. Simultaneous melting of the flux coating on the rod will form gas, and slag, which protects the weld pool from the surrounding atmosphere. This is a versatile process ideal for joining ferrous and non-ferrous materials with a range of material thicknesses in all positions.

Flux Cored Arc Welding (FCAW)

Created as an alternative to SMAW, **FCAW** uses a continuously fed consumable flux cored electrode and a constant voltage power supply, which provides a constant arc length. This process either uses a shielding gas or just the gas created by the flux to provide protection from contamination.

Submerged Arc Welding (SAW)

A frequently-used process with a continuously-fed consumable electrode and a blanket of fusible flux which becomes conductive when molten, providing a current path between the part and the electrode. The flux also helps prevent spatter and sparks while suppressing fumes and ultraviolet radiation.

lectro-Slag Welding (ESW)

A vertical process used to weld thick plates (above 25mm) in a single pass. ESW relies on an electric arc to start before a flux addition extinguishes the arc. The flux melts as the wire consumable is fed into the molten pool, which creates a molten slag on top of the pool. Heat for melting the wire and plate edges is generated through the molten slag's resistance to the passage of the electric current. Two water-cooled copper shoes follow the process progression and prevent any molten slag from running off.

Arc Stud Welding (SW)

Similar to flash welding, **SW** joins a nut or fastener, usually with a flange with nubs that melt to create the join, to another metal piece.

5. NON-CONSUMABLE ELECTRODE METHODS

Tungsten Inert Gas Welding (TIG)

Also known as Gas Tungsten Arc Welding (GTAW), uses a non-consumable tungsten electrode to create the arc and an inert shielding gas to protect the weld and molten pool against atmospheric contamination.

Plasma Arc Welding (PAW)

Similar to TIG, **PAW** uses an electric arc between a non-consumable electrode and an anode, which are placed within the body of the torch. The electric arc is used to ionize the gas in the torch and create the plasma, which is then pushed through a fine bore hole in the anode to reach the base plate. In this way, the plasma is separated from the shielding gas.

6. FORGE WELDING

Forge welding (FOW) is a solid-state welding process that joins two pieces of metal by heating them to a high temperature and then hammering them together. It may also consist of heating and forcing the metals together with presses or other means, creating enough pressure to cause plastic deformation at the weld surfaces. The process is one of the simplest methods of joining metals and has been used since ancient times. Forge welding is versatile, being able to join a host of similar and dissimilar metals. With the invention of electrical and gas welding methods during the Industrial Revolution, manual forge-welding has been largely replaced, although automated forge-welding is a common manufacturing process. Forge welding is a process of joining metals by heating them beyond a certain threshold and forcing them together with enough pressure to cause deformation of the weld surfaces, creating a metallic bond between the atoms of the metals. The pressure required varies, depending on the temperature, strength, and hardness of the alloy. Forge welding is the oldest welding technique, and has been used since ancient times.

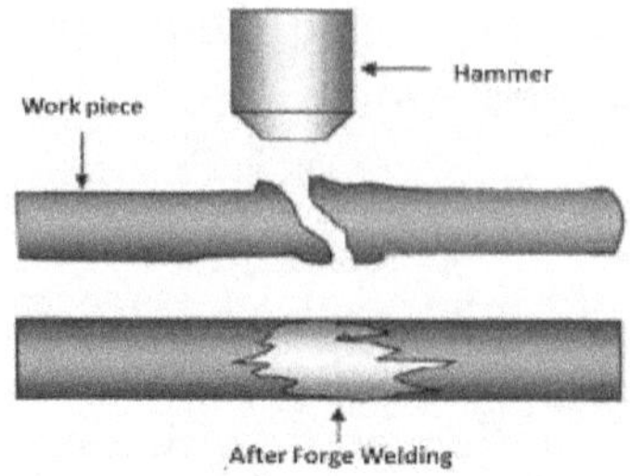

Fig.3. Forge welding

Applications:

- It is used to join steel or iron.
- It is used to manufacture gates, prison cells etc.
- It is widely used in cookware.
- It was used to join boiler plates before introduction of other welding process.
- It was used to weld weapon like sword etc.
- Used to weld shotgun barrels.

Advantages and Disadvantages:

Advantages:

- It is simple and easy.
- It does not require any costly equipment for weld small pieces.
- It can weld both similar and dissimilar metals.
- Properties of weld joint is similar to base material.
- No filler material required.

Disadvantages:

- Only small objects can be weld. Larger objects required large press and heating furnaces, which are not economical.
- High skill required because excessive hammering can damage the welding plates.
- High **Welding defects** involve.
- It cannot use as mass production.
- Mostly suitable for iron and steel.
- It is a slow welding process.

7. SUBMERGED ARC WELDING

The definition of submerged arc welding is, it is one type of welding method where this welding arc can travel under a layer of granular flux. In this type of welding, a tubular electrode otherwise consumable solid can be fed constantly to the weld region. At the same time, a layer of granular fusible flux can be poured over the weld zone which immersed the welding arc as well as defends it from atmospheric pollution.

Equipment of Submerged Arc Welding

The submerged arc welding can be built with main parts or equipment like Welding head, Flux hopper, Flux, Electrode wire feed unit, Electrode, and Flux recovery unit. Welding head can be used to supply filler as well as flux metal to the joint for welding.

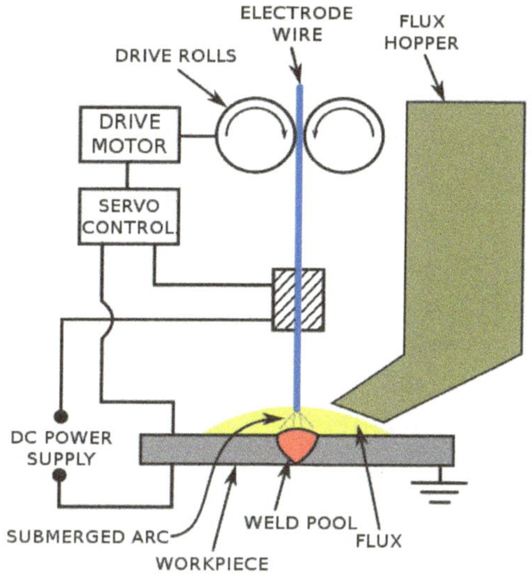

Fig.4. Submerged arc welding

In Flux hopper, the flux can be stored as well as deliver to the welding joint. It controls the rate of deposition of flux to the welding joint.

The granular flux is used to shield the welding arc, and it includes silica, lime, calcium fluoride, oxides of calcium, manganese oxide, etc. It fed into the weld zone with the flow of gravity during the welding head nozzle. Whenever it is melted, then it turns into conductive as well as conducts the current among the workpiece & electrode.

The granular flux's solid layer wraps the melted metal totally & stops the sprinkle and flash. It covers up the UV radiations which is the characteristic of SMAW method. The minor part of the flux obtains melted & shapes slag on the weld pond. It is detached after the welding method obtains completed. The higher element of the flux performs like an insulator & encourages the deep transmission of heat toward the workpiece.

A consumable electrode can be used by the submerged arc welding which is a loop of bare round wire with 1.5 mm to 10 mm diameter. It can be fed routinely throughout the welding gun, and the submerged arc welding electrode composition depends on the welded material. The electrodes are available to weld high carbon steel, mild steel, low and special alloy steels, stainless steel, etc. Generally, the electrodes are covered by the copper to stop rusting & amplify electrical conductivity. They are obtainable within straight length & coils.

Flux recovery unit is used to gather the not used flux present after welding, and after recovery, it can be used another time for the joining.

Submerged Arc Welding Working

In this kind of welding, the flux begins for depositing on the joint to be welded. Whenever the flux is cold, then it acts as an insulator. The arc can be started by moving the tool by the work portion. The arc struck will constantly remain below a wide coating of flux, and the generated heat by the arc softens the granular flux.

Once the flux is melted by the heat of the arc, then it will become highly conductive. The flow of current begins to flow the electrode through the melted flux that can be in contact by the atmosphere. The minor dissolved flux alters to wastage slag & which is detached after welding method finished.

At a fixed speed, the electrode from the roll is constantly fed toward the joint to be linked. If linking is partially automatic, then the top of the welding can be moved physically along with the connection. In an automatic submerged arc welding, a separate drive can be used to move the welding top above the stationary job otherwise job moves beneath the head of the stationary welding.

With the help of the self-adjusting arc principle, the length of the arc is kept stable. When the arc length reduces, the arc voltages will increase & this will increase the arc current.

Because of this, the rates of burn-off will increase & the arc length will be increased. The reverse phenomenon arises when the arc length rises more than the regular length. For straight penetration as well as for supporting the huge quantity of melted metal a support steel plate otherwise copper may be used.

Advantages

The advantages of Submerged Arc Welding include the following.

- This submerged arc welding process has high (45kg/h) deposit rate.
- In automatic applications.
- Very small welding smoke can be observed.
- No edge training is required.
- This method is used in indoor, and outdoor.
- No chance of weld sprinkles because it is submerged within flux blanket.

Disadvantages

The disadvantages of Submerged Arc Welding include the following

- The process is incomplete to some particular metals.
- The application is imperfect to direct seams vessels, and pipes.
- The flux usage is hard.

- A health problem can be occurred due to the flux.
- Slag elimination is desirable after welding.

Submerged Arc Welding Applications

The applications of Submerged Arc Welding include the following

- The Submerged Arc Welding can be used to weld pressure vessels like boilers.
- A lot of structural outlines, pipes, earth moving tools, shipbuilding, railroad construction, and locomotives.
- This type of welding can be used to repair machine parts.

8. RESISTANCE WELDING

Resistance welding can be defined as; it is a liquid state welding method where the metal-to-metal joint can be formed within a liquid state otherwise molten state. This is a thermoelectric method where heat can be generated at the It is a thermo-electric process in which heat is generated at the edge planes of welding plates because of electric resistance and a weld joint can be created by applying low-pressure to these plates. This type of welding uses electric resistance to generate heat. This process is very efficient with pollution free but the applications are limited because of the features like equipment cost is high, and material thickness is limited.

Resistance Welding Working Principle

The working principle of resistance welding is the generation of heat because of electric resistance. The resistance welding such as seam, spot, protection works on the same principle. Whenever the current flows through electric resistance, then heat will be generated. The same working principle can be used within the electric coil. The generated heat will depend on material's resistance, applied current, conditions of a surface, applied the current time period

This heat generation takes place because of the energy conversion from electric to thermal. The resistance welding formula for heat generation is

$$H = I^2RT$$

- 'H' is a generated Heat, and the unit of heat is a joule
- 'I' is an electric current, and the unit of this is ampere
- 'R' is an electric resistance, and the unit of this is Ohm
- 'T' is the time of current flow, and the unit of this is second

The generated heat can be used to soften the edge metal to shape a tough weld joint with fusion. This method generates weld with no application of any flux, filler material, and shielding gases.

Types of Resistance Welding

The following points highlight the five main types of resistance welding. The types are:

1. Spot Welding
2. Projection Welding

3. Seam Welding
4. Butt Welding
5. Flash Butt Welding.

Spot Welding:

It is used for joining two or more sheets of metal. The sheets to be welded are held overlapping between the two electrodes, one moveable and the other fixed. Heavy current is passed through the electrodes. Pressure is applied from the top electrode by moving it downward.

The fusion of material takes place at the spot. The sheet is then moved to have another spot weld at the desired place. The time period for each weld may be 1/100th of a second or more. The current may be of the order of 5000 amperes or more. The voltage between electrodes about 2 volts. The open circuit voltage is less than 12 volts. Rods, wires and small pieces can also be welded by this method. For the sheets the time required is about 1/50sec. for each 0.25 mm of total thickness of the two sheets to be welded.

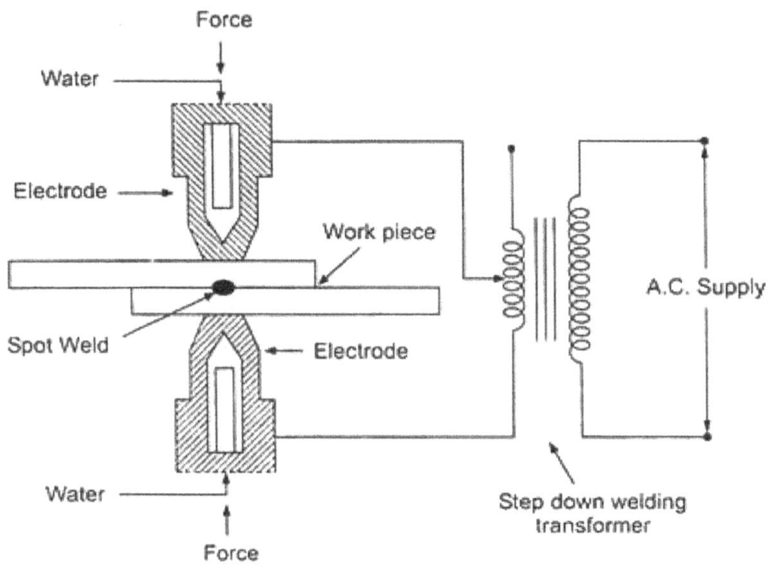

Fig.5. Spot Welding

Advantages of Spot Welding:

(i) It has low initial cost.
(ii) Higher productivity.
(iii) Less skilled worker can operate this welding.
(iv) No edge preparation is required

Disadvantages of Spot Welding:

1. Job of more thickness cannot be welded.
2. High initial cost of equipment.

Applications of Spot Welding:

(i) It can be frequently used in automobile and aircraft industries.
(ii) It is used for making containers.
(iii) Fabrication of steel household furnitures.

Projection Welding:

It is infact a modified form of the spot welding. One of the pieces to be welded this way has projections produced by pressure. The electrodes are flat. The electrodes are placed on work piece and current passed between them. Heat is produced at the contacts and work piece gets welded at these points.

It is easy to weld certain parts which cannot be welded by spot welding. When two plates to be welded are of different cross-section then in order to obtain desirable strength it is necessary to have the projections on the thicker plate.

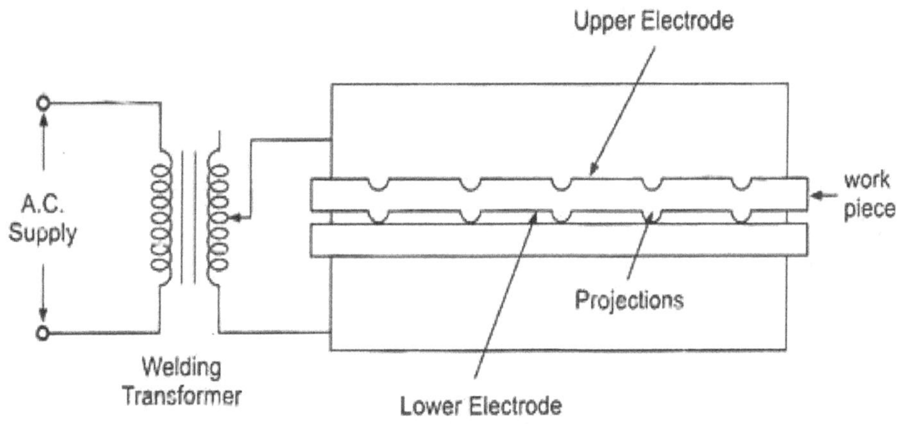

Fig.6. Projection Welding

Advantages of Projection Welding:

(i) More than one spot can be weld at a given time.
(ii) Proper heat balance can be obtained easily.
(iii) The welds may be placed closer than spot welding.
(iv) Electrode life is much longer than the life of electrode in spot welding.
(v) Appearance and uniformity of the weld is better than spot welding.

Disadvantages of Projection Welding:

1. Metals, which cannot support projection, cannot be welded satisfactory.
2. To make projection is an extra operation.
3. The initial cost of the equipment required for projection welding is high as press type machines are required for this type of welding.

Applications of Projection Welding:

(i) It is usually employed on punched where the projection automatically exists.
(ii) Small fasteners, nuts etc. can be welded to larger components.
(iii) Projection welding is used for welding of refrigerator condensers, crossed-wire welding grills etc.

Seam Welding:

In Fig. shows how seam welding is carried out. In this case wheel or roller type electrodes are used and the pressure between them remains constant. The seam welding is done in order to produce a continuous joint which may be required for the construction of tanks, transformer's refrigerators, gasoline tanks, air craft, containers and cylindrical pieces. Seam weld is actually a series of overlapping spot welds. Usually numbers of spots obtained per meter of the joint are between 200 to 400.

The two sheets on which seam welding is required are placed overlapping each other. The current is passed between the two moving electrodes as shown in Fig. 4.7. The electrodes are to be shape of continuously rotating wheel which also applies pressure on the job. It is used when a continuous joint required between two overlapping metal sheets.

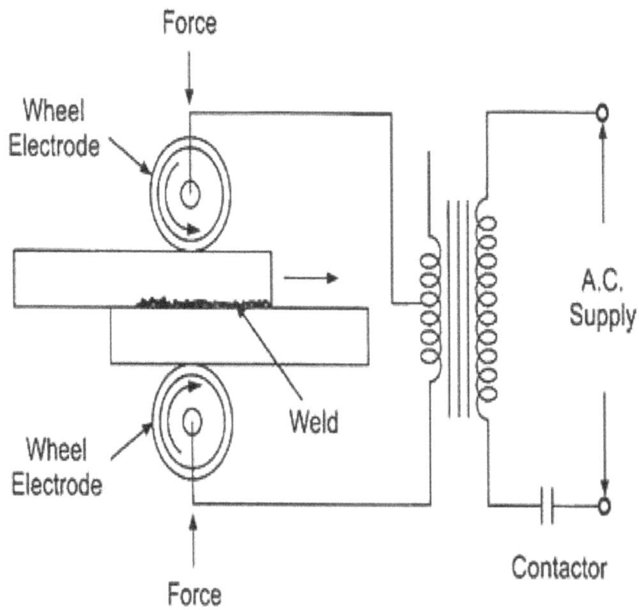

Fig.7. Seam Welding

Advantages of Seam Welding:

(i) Less overlap than spot or projection welds.
(ii) Several parallel seams may be produced.
(iii) Gas tight or liquid tight joints can be produced.

Disadvantages of Seam Welding:

(i) High cost of equipment as compared to spot welding set.
(ii) It is difficult to weld thickness greater than 3 mm.

Applications of Seam Welding:

1. It is used for welding of stainless steel, aluminium and its alloys nickel and its alloys etc.

Butt Welding:

In Fig.4.8 shows how butt welding is done. The welding transformer has a comparatively larger number of turns on the primary and lesser number of turns on the secondary because heavy current is required on the secondary side. The two parts to be welded together are placed touching each other end to end so as to form a butt joint.

A pressure is also applied in the axial direction of weld. The jobs are securely clamped and heavy current is passed through them. The heat produced is sufficient to raise the temperature of material to plastic state and fusion takes place at the points of contact when pressure is applied on both sides.

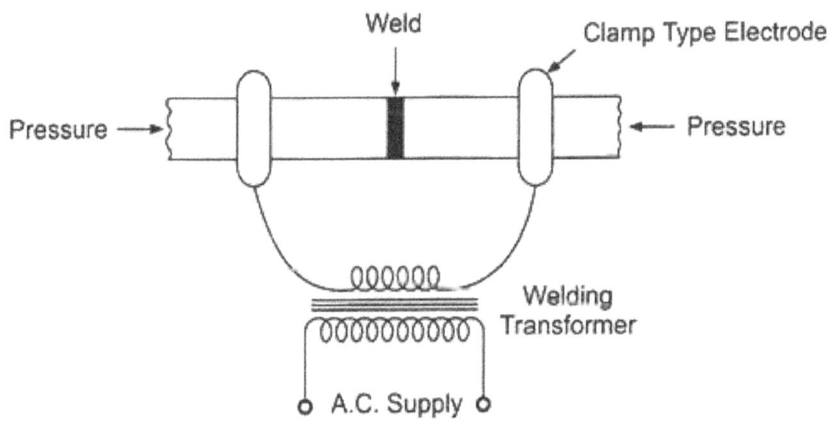

Fig.8.Butt Welding

This method is used for welding pipes, wires and rods etc. The voltage required for welding is 2 to 10 volts and current varies from 50 A to several hundred amperes, depending upon the material and the area to be welded at a time.

Flash Butt Welding:

This is similar to butt welding except that the parts to be welded are joined together under light pressure and a heavy current passed through the joint. Due to poor contact at the joint arcing takes place. When sufficient heat has been produced the two parts are suddenly pressed together and the current is simultaneously stopped. A thin film is produced around the joint which is subsequently removed to give a weld of this type. It is used for welding chains, rail ends, shaft axles etc.

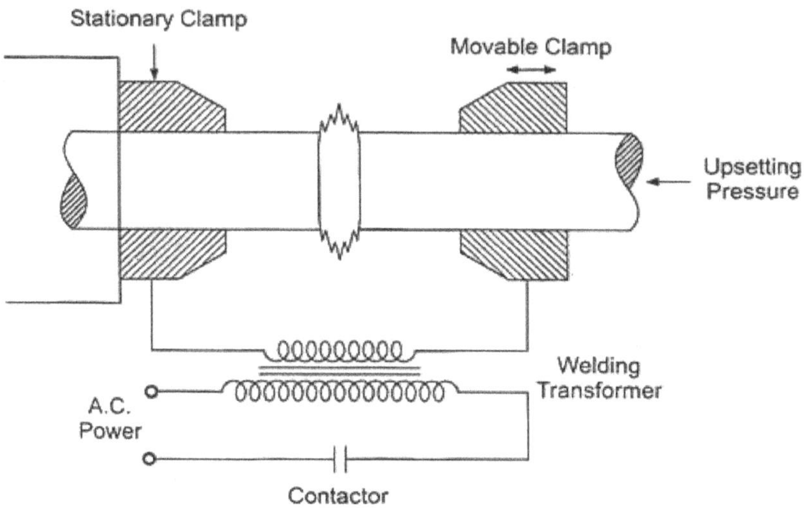

Fig.9. Flash Butt Welding

Advantages of Flash Butt Welding:

1. It is a faster process.
2. Preparation of weld surface is not required.
3. Power requirement is less.
4. Many dissimilar metals with different melting temperatures can be flash welded.
5. The process is cheap.
6. It offers strength factor upto 100%.

Disadvantages of Flash Butt Welding:

1. More chances of fire hazards.
2. Metal is lost during flashing and upsetting.
3. Concentricity and straightness of the work pieces during welding process is often difficult to maintain.

Applications of Flash Butt Welding:

1. It is used for welding metal sheets, bars, rods fittings etc.
2. Flash welding finds applications in automotive and air craft products.
3. It is also employed in household appliances, refrigerators and farm implements.

9. THERMIT WELDING

Thermit welding is a chemical welding process in which an exothermic chemical reaction is used to supply the essential heat energy. That reaction involves the burning of Thermit, which is a mixture of fine aluminum powder and iron oxide in the ratio of about 1:3 by weight.

Although a temperature of 3000°C may be attained as a result of the reaction, preheating of the Thermit mixture up to about 1300°C is essential in order to start the reaction.

The mixture reacts according to the chemical reaction:

$$8\,Al + 3\,Fe_3O_4 \rightarrow 9\,Fe + 4\,Al_2O_3 + \text{heat } (3000^\circ C,\ 35\ kJ/kg \text{ of mixture})$$

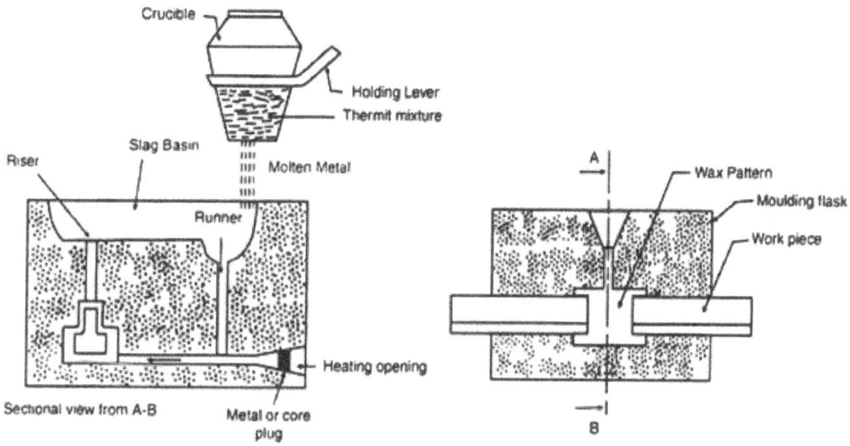

Fig.10.Thermit Welding

Aluminum has greater affinity to react with oxygen; it reacts with ferric oxide to liberate pure iron and slag of aluminum oxide. Aluminum oxide floats on top of molten metal pool in the form of slag and pure iron (steel) settled below, because of large difference in densities.

Operation of Thermit Welding:

Thermit welding process is essentially a casting and foundry process, where the metal obtained by the Thermit reaction is poured into the refractory cavity made around the joint.

The various steps involved in Thermit welding are:

1. The two pieces of metal to be joined are properly cleaned and the edge is prepared.
2. Then the wax is poured into the joint so that a wax pattern is formed where the weld is to be obtained.

3. Amolding box is kept around the joint and refractory sand is packed carefully around the wax pattern as shown in Fig., providing the necessary pouring basin, sprue, and riser and gating system.
4. A bottom opening is provided to run off the molten wax. The wax is melted through this opening which is also used to preheat the joint. This makes it ready for welding.
5. The Thermit is mixed in a crucible which is made of refractory material that can withstand the extreme high heat and pressure, produced during the chemical reaction.
6. The igniter (normally barium peroxide or magnesium) is placed on top of the mixture and is lighted with a red hot metal rod or magnesium ribbon.
7. The reaction takes about 30 seconds and highly super-heated molten iron is allowed to flow into the prepared mould cavity around the part to be welded.
8. The super-heated molten metal fuses the parent metal and solidifies into a strong homogeneous weld.
9. The weld joint is allowed to cool slowly.

There are different Thermit mixtures available for welding different metals, such as copper and chromium. They use different metal oxides in place of ferrous oxide. Some typical Thermit mixture reactions with their temperature obtained are given below:

$3\,CuO + 2\,Al \rightarrow 3Cu + Al_2O_3$ + Heat (4860°C, 275 Kcal)

$Cr_2O_3 + 2Al \rightarrow 2Cr + Al_2O_3$ + Heat (3000°C, 540 Kcal)

Application and Uses of Thermit Welding:

Thermit welding is a very old process and now-a-days, in most cases, it is replaced by electro-slag welding. However, this process is still in use.

Some applications are:

(i) Thermit welding is traditionally used for the welding of very thick and heavy plates.
(ii) Thermit welding is used in joining rail roads, pipes and thick steel sections.
(iii) Thermit welding is also used in repairing heavy castings and gears.
(iv) Thermit welding is suitable to weld large sections such as locomotive rails, ship hulls etc.
(v) Thermit welding is used for welding cables made of copper.

Advantages of Thermit Welding:

1. Thermit welding is a simple and fast process of joining similar or dissimilar metals.
2. This process is cheap, as no costly power supply is required.
3. This process can be used at the places where power supply is not available.

Disadvantages of Thermit Welding:

1. Thermit welding is essentially used for ferrous metal parts of heavy sections.
2. It is uneconomical for welding cheap metals and light parts.

UNIT - III

ADVANCES IN WELDING PROCESS

1. METAL INERT GAS WELDING

Metal Inert Gas (MIG) welding is a gas metal arc welding (GMAW) process that uses a continuous solid wire electrode which is heated and fed into the weld pool from a welding gun. The two base materials are melted together which causes them to join. The welding gun also feeds an inert shielding gas alongside the wire electrode, which helps protect the process from airborne contaminants.

Process Characteristics

MIG/MAG welding is a versatile technique suitable for both thin sheet and thick section components. An arc is struck between the end of a wire electrode and the workpiece, melting both of them to form a weld pool. The wire serves as both heat source (via the arc at the wire tip) and filler metal for the joint. The wire is fed through a copper contact tube (contact tip) which conducts welding current into the wire. The weld pool is protected from the surrounding atmosphere by a shielding gas fed through a nozzle surrounding the wire. Shielding gas selection depends on the material being welded and the application. The wire is fed from a reel by a motor drive, and the welder moves the welding torch along the joint line. Wires may be solid (simple drawn wires), or cored (composites formed from a metal sheath with a powdered flux or metal filling). Consumables are generally competitively priced compared with those for other processes. The process offers high productivity, as the wire is continuously fed.

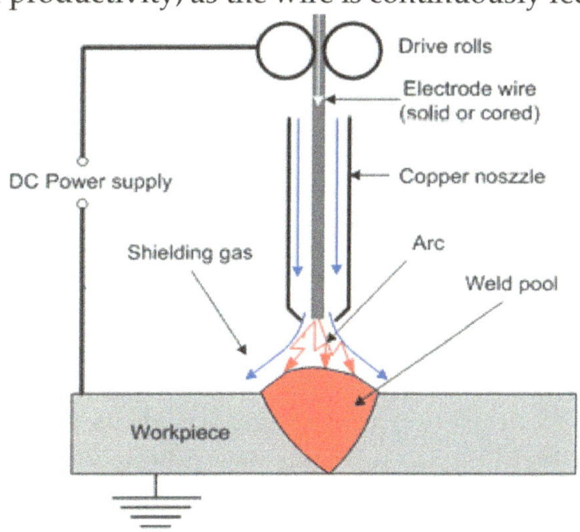

Fig.1.Metal Inert Gas Welding

Metal Transfer Mode

The manner, or mode, in which the metal transfers from the electrode to the weld pool largely determines the operating features of the process. There are three principal metal transfer modes:

- Short circuiting/ Dip
- Droplet / spray
- Pulsed

Short-circuiting and pulsed metal transfer are used for low current operation while spray metal transfer is only used with high welding currents. In short-circuiting or 'dip' transfer, the molten metal forming on the tip of the wire is transferred by the wire dipping into the weld pool. This is achieved by setting a low voltage; for a 1.2mm diameter wire, arc voltage varies from about 17V (100A) to 22V (200A). Care in setting the voltage and the inductance in relation to the wire feed speed is essential to minimize spatter. Inductance is used to control the surge in current which occurs when the wire dips into the weld pool.

For droplet or spray transfer, a much higher voltage is necessary to ensure that the wire does not make contact i.e. short-circuit, with the weld pool; for a 1.2mm diameter wire, the arc voltage varies from approximately 27V (250A) to 35V (400A). The molten metal at the tip of the wire transfers to the weld pool in the form of a spray of small droplets (about the diameter of the wire and smaller). However, there is a minimum current level, threshold, below which droplets are not forcibly projected across the arc. If an open arc technique is attempted much below the threshold current level, the low arc forces would be insufficient to prevent large droplets forming at the tip of the wire. These droplets would transfer erratically across the arc under normal gravitational forces. The pulsed mode was developed as a means of stabilizing the open arc at low current levels i.e. below the threshold level, to avoid short-circuiting and spatter. Metal transfer is achieved by applying pulses of current, each pulse having sufficient force to detach a droplet. Synergic pulsed MIG refers to a special type of controller which enables the power source to be tuned (pulse parameters) for the wire composition and diameter, and the pulse frequency to be set according to the wire feed speed.

Shielding Gas

In addition to general shielding of the arc and the weld pool, the shielding gas performs a number of important functions:

- Forms the arc plasma
- Stabilizes the arc roots on the material surface
- Ensures smooth transfer of molten droplets from the wire to the weld pool

Thus, the shielding gas will have a substantial effect on the stability of the arc and metal transfer and the behavior of the weld pool, in particular, its penetration. General purpose shielding gases for MIG welding are mixtures of argon, oxygen and CO_2, and special gas mixtures may contain helium. The gases which are normally used for the various materials are:

Steels:

- CO_2
- Argon +2 to 5% oxygen
- Argon +5 to 25% CO_2

Non-ferrous (e.g. Aluminium, copper or nickel alloys):

- Argon
- Argon / helium

Argon based gases, compared with CO_2, are generally more tolerant to parameter settings and generate lower spatter levels with the dip transfer mode. However, there is a greater risk of lack of fusion defects because these gases are colder. As CO_2 cannot be used in the open arc (pulsed or spray transfer) modes due to high back-plasma forces, argon based gases containing oxygen or CO_2 are normally employed.

Applications:

- MIG is best suited for fabrication of sheet metal.
- Generally all available metals can be weld through this process.
- It can be used for deep groove welding.

Advantages:

- It provide higher deposition rate.
- It is faster comparing to arc welding because it supply filler material continuously.
- It produce clean weld with better quality.
- There is no slag formation.
- Minimize weld defects.
- This welding produces very little slag.
- It can be used to make deep groove weld.
- It can be easily automated.

Disadvantages:

- It cannot be used for welding in difficult to reach portions.
- Higher initial or setup cost.
- It cannot be used for outdoor work because wind can cause damage of gas shield.
- It required high skilled labor.

2. TUNGSTEN INERT GAS WELDING (TIG WELDING)

Tungsten Inert Gas Welding (TIG Welding), also known as Gas Tungsten Arc Welding (GTAW), is an arc welding method that uses a non-consumable tungsten electrode to weld two or more workpieces. It is very much similar to Metal Inert Gas (MIG) Welding.

Components used in Tungsten Inert Gas Welding (TIG Welding):

The following components are necessary to perform Tungsten Inert Gas Welding:

1. Power Supply (A.C or D.C)
2. Non-consumable Tungsten Electrode

3. Inert Gas Supply
4. Filler Rod (used depending on the nature of workpiece)
5. Welding Head

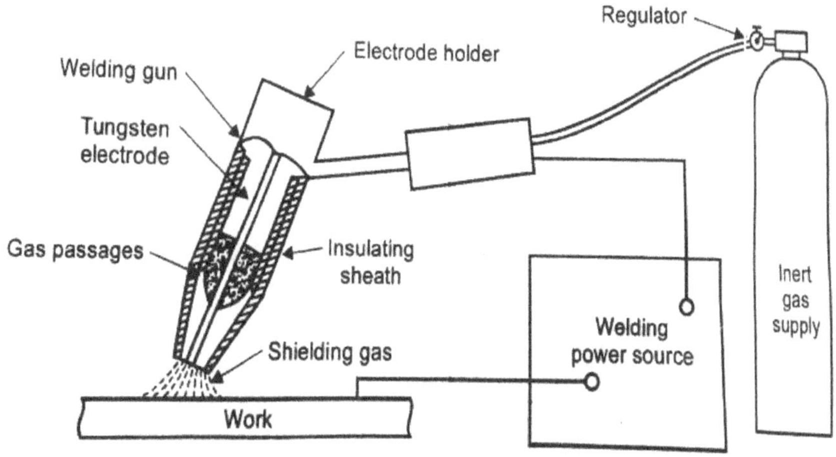

Fig.2.Tungsten Inert Gas Welding

Working principle:

The workpiece to be welded is placed on the worktable. The non-consumable tungsten electrode and the workpiece are connected to the power supply (A.C or D.C).As the electrode is brought near the workpiece (leaving a small air gap), an arc is produced. This arc is used for melting and welding the workpiece.Tungsten has high melting point (3422 °C). Hence, tungsten electrode does not melt during the welding process.

In tungsten inert gas welding, filler rod may or may not be used. The usage of filler rod depends on the nature of the workpiece to be welded. If filler rod is used, it is continuously melted by the arc and fed into the weld pool.

Inert gas supply is constantly provided around the electrode during the welding process. The inert gas forms a gas shielding around the weld. It protects the weld from the external atmosphere. (See diagram above for better understanding).

Advantages of Tungsten Inert Gas Welding (TIG Welding):

1. Tungsten Inert Gas Welding produces high quality welds.
2. The weld is automatically protected by the inert gas during the welding process.
3. No slag is produced.
4. TIG Welding can be done in any position.

Disadvantages of Tungsten Inert Gas Welding:

1. Tungsten inert gas welding is a slow process.
2. Highly skilled labor is needed.
3. Welder is exposed to huge intensities of light.
4. TIG welding is more expensive when compared to MIG welding.

Applications of Tungsten Inert Gas Welding:

TIG Welding is used for welding a variety of metals. Some of them are:

1. Stainless steel
2. Alloy steel
3. Aluminium
4. Titanium
5. copper
6. magnesium
7. nickel alloys

3. FRICTION WELDING

Friction welding works on basic principle of friction. In this welding process, the friction is used to generate heat at the interference surface. This heat is further used to join two work pieces by applying external pressure at the surface of work piece. In this welding process, the friction is applied until the plastic forming temperature is achieved. It is normally 900-1300 degree centigrade for steel. After this heating phase, a uniformly increasing pressure force applied until the both metal work pieces makes a permanent joint. This joint is created due to thermo mechanical treatment at the contact surface.

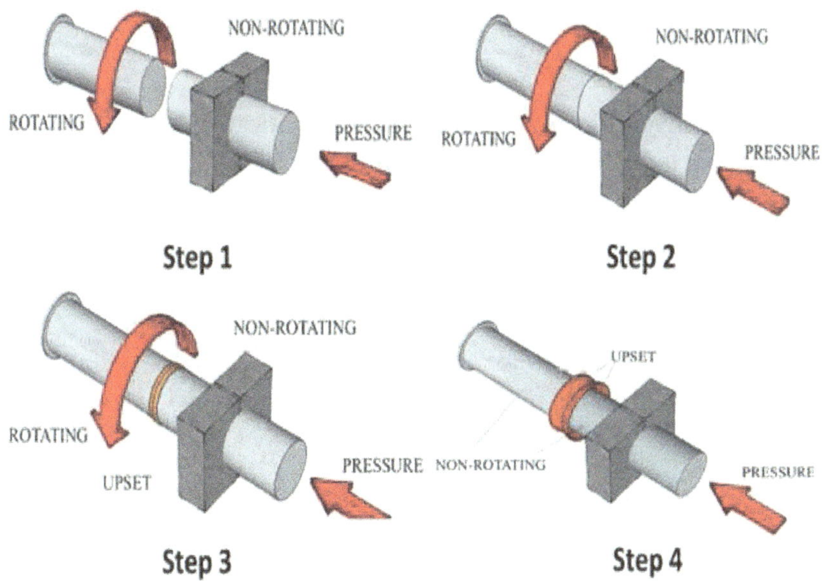

Fig.3. Friction welding

Working principle:

- First both the work pieces are prepared for smooth square surface. One of them is mounted on a rotor driven chuck and other one remains stationary.

- The rotor allows rotating at high speed thus it makes rotate mounted work piece. A little pressure force is applied on the stationary work piece which permits cleaning the surface by burnishing action.
- Now a high pressure force applied to the stationary work piece which forces it toward rotating work piece and generates a high friction force. This friction generates heat at the contact surface. It is applied until the plastic forming temperature is achieved.
- When the temperature is reached the desire limit, the rotor is stopped and the pressure force is applied increasingly until the whole weld is formed.
- This welding is used to weld those metals and alloys which cannot be welded by other method.

Applications:

- For welding tubes and shafts.
- It is mostly used in aerospace, automobile, marine and oil industries.
- Gears, axle tube, valves, drive line etc. components are friction welded.
- It is used to replace forging or casting assembly.
- Hydraulic piston rod, truck rollers bushes etc. are join by friction welding.
- Used in electrical industries for welding copper and aluminum equipment's.
- Used in pump for welding pump shaft (stainless steel to carbon steels).
- Gear levers, drill bits, connecting rod etc. are welded by friction welding.

Advantages:

- It is environment friendly process without generation smoke etc.
- Narrow heat affected zone so no change in properties of heat sensitive material.
- No filler metal required.
- Welding strength is strong in most cases.
- Easily automated.
- High welding speed.
- High efficiency of weld.
- Wide variety of metal can be weld by this process.

Disadvantages:

- This is mostly used only for round bars of same cross section.
- Non-forgeable material cannot be weld.
- Preparation of work piece is more critical
- High setup cost.
- Joint design is limited.

4. INDUCTION WELDING

Induction welding is a form of welding that uses electromagnetic induction to heat the workpiece. The welding apparatus contains an induction coil that is energized with a radio-frequencyelectric current. This generates a high-frequency electromagnetic field that acts on either an electrically conductive or a ferromagnetic workpiece. In an electrically conductive workpiece, the main heating effect is resistive heating, which is due to induced currents called eddy currents. In a ferromagnetic workpiece, the heating is caused mainly by hysteresis, as the electromagnetic field repeatedly distorts the magnetic domains of the ferromagnetic material. In practice, most materials undergo a combination of these two effects.

Nonmagnetic materials and electrical insulators such as plastics can be induction-welded by implanting them with metallic or ferromagnetic compounds, called susceptors, that absorb the electromagnetic energy from the induction coil, become hot, and lose their heat to the surrounding material by thermal conduction. [1] Plastic can also be induction welded by embedding the plastic with electrically conductive fibers like metals or carbon fiber. Induced eddy currents resistively heat the embedded fibers which lose their heat to the surrounding plastic by conduction. Induction welding of carbon fiber reinforced plastics is commonly used in the aerospace industry.

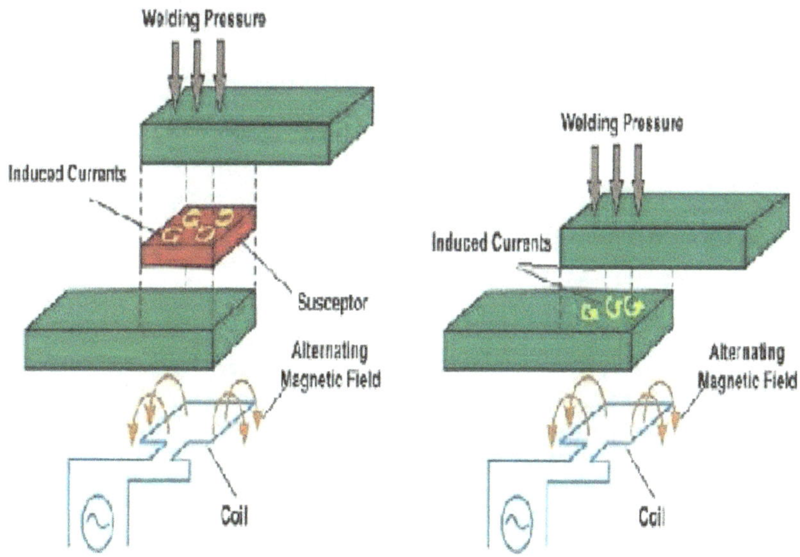

Fig.4. Induction welding

Induction welding is used for long production runs and is a highly automated process, usually used for welding the seams of pipes. It can be a very fast process, as a lot of power can be transferred to a localized area, so the faying surfaces melt very quickly and can be pressed together to form a continuous rolling weld.

The depth that the currents, and therefore heating, penetrates from the surface is inversely proportional to the square root of the frequency. The temperature of the metals being welded and their composition will also affect the penetration depth. This process is very similar to resistance welding, except that in the case of resistance welding the current is delivered using contacts to the workpiece instead of using induction.

When induction welding takes place, the work pieces heat up to under the melting temperature and the edges of the pieces are placed together impurities get forced out to give a solid forge weld. Induction welding is used for joining a multitude of thermoplastics and thermosetting matrix composites. The apparatus used for induction welding processes includes a radio frequency power generator, a heating station, the work piece material, and a cooling system.

The power generator comes in either the form of solid state or vacuum tube and is used to provide an alternating current of 230-340 V or a frequency of 50-60 Hz to the system. This value is determined by what induction coil is used with the piece. The heat station utilizes a capacitor and a coil to heat the work pieces. The capacitor matches the power generators output and the induction coil transfers energy to the piece. When welding the coil needs to be close to the work piece to maximize the energy transfer and the work piece used during induction welding is an important key component of optimal efficiency.

Advantages:

- Less effective splitting;
- High current density and high efficiency due to the proximity of the contact head;
- Small thermal influence range;
- Small influence of pipe diameter, weldable pipes with a diameter of about 8 to 1270 mm;
- Welding speeds of up to 100 to 120 m/min

Disadvantages:

- Consumption of the contact head;
- Erosion defects;
- Small heating range, rapid burning, which is unfavorable to the inner surface of the tube.

5. EXPLOSION WELDING

Explosion welding, also called explosive cladding, joins two metals together by an explosive force. It is often used to join dissimilar metals that have a significant difference in mechanical properties, thermal expansion coefficients, or melting temperatures – such as a low-cost metal to a more expensive corrosion-resistant metal. Some of the common clad layers deposited onto steel plate are aluminum, copper, bronze, titanium, monel, nickel alloys and zirconium.

An explosive weld is made by impelling a cladding plate against a substrate plate using significant energy from an explosive discharge, which results in a high energy rate impact. Very high forces at the impact area cause the first few atomic layers of each material to form a plasma jet that is ejected out of the impact zone, taking with it any contamination and unwanted oxide layers. The impact causes the metals to form a true metallic bond, where the metals share valence electrons.

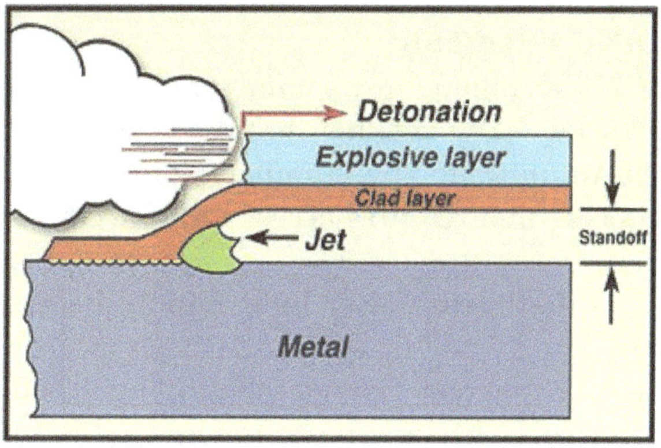

Fig.5. Explosion welding

Explosion welding traditionally produces wavy bond zone morphology. This waviness is a function of the different material properties and the welding parameters used. Impact angle and the detonation velocity influence the amplitude of the waviness.When properly made, an explosion weld is a solid-state bond without any heat affected zone to degrade its strength, as is typical of other heat-dependent welding or joining processes.Explosion welding is well-suited for joining large, flat areas, as in clad plate manufacture. To get more complex components, a bimetal perform is typically produced by explosion welding in the plate or tube configuration, followed by hot or cold working.The explosive welding technique is frequently used to clad low cost plate (usually carbon steel) with more expensive corrosion resistant materials. This clad plate is typically used as tube sheet for heat exchangers in the chemical and petrochemical industries. Other applications include sandwiched metal for coinage and the production of titanium-to-stainless steel transition joints in the Apollo spacecraft.

A broad range of sizes can be explosively welded; it has been used to join electronic components of widths less than 0.5 mm and for cladding plates up to 5 x 12 m. Cladding metal thicknesses can range from 0.025 to 100 mm (0.001 to 4 in.), and base-metal thickness can range from 0.025 mm (0.001 in.) to over 1 m (40 in.). The size limits are generally mandated by component metal manufacturability and transportation requirements. Because of the unique safety and noise-vibration considerations, explosion welding is usually performed in isolated facilities by experienced companies.

Advantages of explosion welding

- Large areas can be bonded quickly
- Stronger bonds than friction and diffusion welded joints

Limitations

- Requires knowledge of explosives
- May require special licensing for high explosives
- Limited to flat surfaces or coaxial cylindrical surfaces

6. LASER BEAM WELDING PROCESS

Laser beam welding is a technique in manufacturing whereby two or more pieces of material (usually metal) are joined by together through use of a laser beam.

Laser stands for **L**ight **A**mplification by **S**timulated **E**mission of **R**adiation.

It is a non-contact process that requires access to the weld zone from one side of the parts being welded.

The weld is formed as the intense laser light rapidly heats the material - typically calculated in milliseconds.

The laser beam is a coherent (single-phase) light of a single wavelength (monochromatic). The laser beam has low beam divergence and high energy content and thus will create heat when it strikes a surface

The primary types of lasers used in welding and cutting are:

- **Gas lasers**: use a mixture of gases such as helium and nitrogen. There are also CO2 or carbon dioxide lasers. These lasers use a low-current, high-voltage power source to excite the gas mixture using a lasing medium. Operate in a pulsed or continuous mode.

  Carbon dioxide lasers use a mixture of high purity carbon dioxide with helium and nitrogen as the lasing medium. CO2 lasers are also used in dual-beam laser welding where the beam is split into two equal power beams.
- **Solid state lasers**: (Nd:YAG type and ruby lasers) Operate at 1micrometer wavelengths. They can be pulsed or operate continuously. Pulsed operation produced joints similar to spot welds but with complete penetration. The pulse energy is 1 to 100 Joules. Pulse time is 1 to 10 milliseconds.
- **Diode lasers**:Lasers are used for materials that are difficult to weld using other methods, for hard to access areas and for extremely small components. Inert gas shielding is needed for more reactive materials.

Laser beam welding (LBW) is a welding process that produces coalescence of materials with the heat obtained from the application of a concentrate coherent light beam impinging upon the surfaces to be joined.

The focused laser beam has the highest energy concentration of any known source of energy. The laser beam is a source of electromagnetic energy or light that can be pro jetted without diverging and can be concentrated to a precise spot. The beam is coherent and of a single frequency.

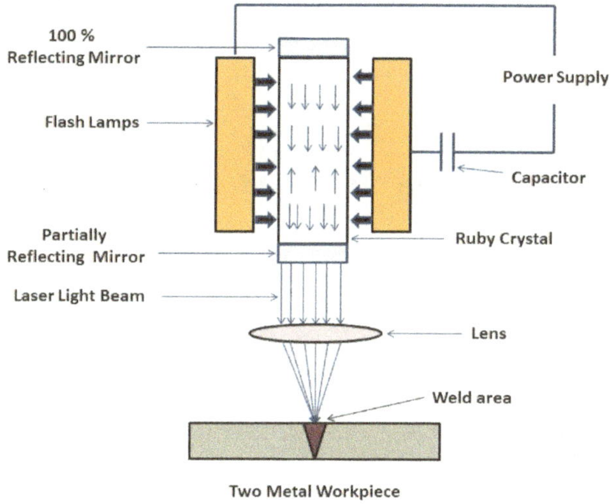

Fig.6. Laser beam welding

Working principle

- Pump source provides energy to the medium, exciting the laser such that electrons held within the atoms are elevated temporarily to higher energy states.
- The electrons held in this excited state cannot remain there indefinitely and drop down to a lower energy level.
- The electron looses the excess energy gained from the pump energy by emitting a photon. This is called spontaneous emission and the photons produced by this method are the seed for laser generation.
- Photons emitted by spontaneous emission eventually strike other electrons in the higher energy states. The incoming photon "knocks" the electron from the excited state to a lower energy level creating another photon. These photons are coherent meaning they are in phase, of the same wavelength, and traveling the same direction. A process called stimulated emission.
- Photons are emitted in all directions, however some travel along the laser medium to strike the resonator mirrors to be reflected back through the medium. The resonator mirrors define the preferential amplification direction for stimulated emission. In order for the amplification to occur there must be a greater percentage of atoms in the excited state than the lower energy levels. This population inversion of more atoms in the excited state leads to the conditions required for laser generation.
- The focus spot of the laser is targeted on the workpiece surface which will be welded. At the surface the concentration of light energy converts into thermal energy (heat). The heat causes the surface of the material to melt, which progresses through the surface by a process called surface conductivity. The beam energy level is maintained below the vaporization temperature of the workpiece material.The

ideal thickness of the materials to be welded is 20mm. The energy is a laser is concentrated, an advantage when working with materials that have high thermal conductivity.

Metals Used in Laser Welding

The laser beam has been used to weld:

- Carbon Steel
- Aluminum
- Titanium
- Low alloy and stainless steel
- Nickel
- Platinum
- Molybdenum

Laser welds made in these materials are similar in quality to welds made in the same materials by electron beam process.

Experimental work using filler metal is being used to weld metals that tend to show porosity when welded with either EB or LB welding. Materials 1/2 in. (12.7 mm) thick are being welded at a speed of 10.0 in. (254.0 mm) per minute.

Advantages of Laser Welding

- Works with high alloy metals without difficulty
- Can be used in open air
- Can be transmitted over long distances with a minimal loss of power
- Narrow heat affected zone
- Low total thermal input
- Welds dissimilar metals
- No filler metals necessary
- No secondary finishing necessary
- Extremely accurate
- Produces deep and narrow welds
- Low distortion in welds
- High quality welds
- Can weld small, thin components
- No contact with materials

Limitations

- Rapid cooling rate may cause cracking in some metals
- High capital cost for equipment
- Optical surfaces of the laser are easily damaged
- High maintenance costs

Soldering and brazing jointly represent one of several methods used for joining two or more pieces of metal. In essence, a joint is made in metal using an alloy of two or more metals to hot-glue the parts together. The word "glue" here does not mean just sticking something together as the items to be joined are to be bonded at a molecular level to the alloy, which imparts considerable strength to the joint. Soldering and brazing are the only metal joining methods that can produce smooth and rounded fillets at the periphery of the joints. Both the operations involve heating the filler metal and joint surfaces above ambient temperature. Both are essentially the same metal joining technique, the difference being the temperature at which each method is performed. Brazing filler metals melt above 450 °C. The solder filler metals melt below 450 °C.

7. SOLDERING PROCESS

Soldering is one of the oldest and most popular techniques used for joining similar or dissimilar metals. It uses a filler material to join parent materials that remain solid. It is a low-temperature analog to brazing which uses filler alloys with melting temperatures below 450 °C (840 °F). This process may or may not require a fluxing agent. The filler metals melt at low temperatures so there is minimum part distortion and heat damage to sensitive parts. The filler metal is called solder which when solidifies, is then bonded to the metal parts to join them. The most commonly used solder is tin alloy and lead. Soldering is extensively used in the electronics industry for joining wires, capacitor, resistor, etc. with the joining plate.

Solders

Soldering filler materials are available in many different alloys for differing applications. In electronics assembly, the eutectic alloy with 63% tin and 37% lead (or 60/40, which is almost identical in melting point) has been the alloy of choice. Other alloys are used for plumbing, mechanical assembly, and other applications. Some examples of soft-solder are tin-lead for general purposes, tin-zinc for joining aluminium, lead-silver for strength at higher than room temperature, cadmium-silver for strength at high temperatures, zinc-aluminium for aluminium and corrosion resistance, and tin-silver and tin-bismuth for electronics.

Flux

The purpose of flux is to facilitate the soldering process. One of the obstacles to a successful solder joint is an impurity at the site of the joint, for example, dirt, oil or oxidation. The impurities can be removed by mechanical cleaning or by chemical means, but the elevated temperatures required to melt the filler metal (the solder) encourages the work piece (and the solder) to re-oxidize. This effect is accelerated as the soldering temperatures increase and can completely prevent the solder from joining to the workpiece. One of the earliest forms of flux was charcoal, which acts as a reducing agent and helps prevent oxidation during the soldering process. Some fluxes go beyond the simple prevention of oxidation and also provide some form of chemical cleaning (corrosion). Many fluxes also act as a wetting agent in the soldering process, reducing the surface tension of the molten solder and causing it to flow and wet the workpieces more easily.

There are three forms of soldering, each requiring progressively higher temperatures and producing an increasingly stronger joint strength:

1. soft soldering, which originally used a tin-lead alloy as the filler metal
2. silver soldering, which uses an alloy containing silver
3. brazing which uses a brass alloy for the filler

Soft soldering is characterized by having a melting point of the filler metal below approximately 400 °C (752 °F),whereas silver soldering and brazing use higher temperatures, typically requiring a flame or carbon arc torch to achieve the melting of the filler. Soft solder filler metals are typically alloys (often containing lead) that have liquidus temperatures below 350 °C (662 °F).

In this soldering process, heat is applied to the parts to be joined, causing the solder to melt and to bond to the workpieces in a surface alloying process called wetting. In stranded wire, the solder is drawn up into the wire between the strands by capillary action in a process called 'wicking'. Capillary action also takes place when the workpieces are very close together or touching. The joint's tensile strength is dependent on the filler metal used; in electrical soldering little tensile strength comes from the added solder which is why it is advised that wires be twisted or folded together before soldering to provide some mechanical strength for a joint. A good solder joint produces an electrically-conductive, water- and gas-tight join.

Silver soldering

"Hard soldering" or "silver soldering" is used to join precious and semi-precious metals such as gold, silver, brass, and copper. The solder is usually described as easy, medium, or hard in reference to its melting temperature, not the strength of the joint. Extra-easy solder contains 56% silver and has a melting point of 618 °C (1,145 °F). Extra-hard solder has 80% silver and melts at 740 °C (1,370 °F). If multiple joints are needed, then the jeweler will start with hard or extra-hard solder and switch to lower-temperature solders for later joints.

Silver solder is somewhat absorbed by the surrounding metal, resulting in a joint that is actually stronger than the metal being joined. The metal being joined must be perfectly flush, as silver solder cannot normally be used as a filler and will not fill gaps.

Induction soldering

Induction soldering uses induction heating by high-frequency alternating current in a surrounding copper coil. This induces currents in the part being soldered, which generates heat because of the higher resistance of a joint versus its surrounding metal (resistive heating). These copper coils can be shaped to fit the joint more precisely. A filler metal (solder) is placed between the facing surfaces, and this solder melts at a fairly low temperature. Fluxes are commonly used in induction soldering. This technique is particularly suited to continuously soldering, in which case these coils wrap around a cylinder or a pipe that needs to be soldered.

Applications

- Soldering is used in plumbing, electronics, and metalwork from flashing to jewellery and musical instruments.
- Soldering provides reasonably permanent but reversible connections between copper pipes in plumbing systems as well as joints in sheet metal objects such as food cans, roof flashing, rain gutters and automobile radiators.
- Jewelry components, machine tools and some refrigeration and plumbing components are often assembled and repaired by the higher temperature silver soldering process. Small mechanical parts are often soldered or brazed as well. Soldering is also used to join lead came and copper foil in stained glass work.
- Electronic soldering connects electrical wiring and electronic components to printed circuit boards by utilizing a metallic alloy substance called solder. This special alloy is melted by using a soldering iron, a wave bath, or a specialized oven, as it joins conductors to PCBs, terminals, and wires.
- Musical instruments, especially brass and woodwind instruments use a combination of soldering and brazing in their assembly. Brass bodies are often soldered together, while keywork and braces are most often brazed.

8. BRAZING PROCESS

Brazing is yet another metal joining process in which two or metals are joined together by heating and melting a filler metal, which then bind the two pieces together and join them. It can be applied to a wide variety of materials, including metals, ceramics, glasses, plastic, and composite materials. Although it is not as strong as fusion welding, it is the strongest form of metal bonding without melting the parent metal of the components being joined together. So this process requires more heat input than other soldering operations such as mechanical fastening, adhesive bonding, solid state joining, welding and so on. This process can also be used to join dissimilar metals such as silver, gold, copper, aluminum, etc. Brazing is necessarily done at temperature above 450 °C but below the critical temperature of metal.

Localized Heating Techniques

- **Torch Brazing**

  In this method, the heat required to melt and flow filler metal is supplied by a fuel gas flame. The fuel gas can be acetylene, hydrogen, or propane and is combined with oxygen or air to form a flame. This process is readily automated and requires low capital investment. Torch brazing requires the use of a flux, so a post-braze clean is often required.

- **Induction Brazing**
  High frequency induction heating for brazing is clean and rapid, giving close control of temperature and location of heat. Heat is created by a rapidly alternating current which is induced into the workpiece by an adjacent coil.
- **Resistance Brazing**
  This is a process in which heat is generated from resistance to an electrical current (as for induction brazing) flowing in a circuit which includes the workpieces. The process is most applicable to relatively simple joints in metals which have high electrical conductivity.

Diffuse Heating Techniques

- **Furnace Brazing**
  Furnace brazing offers two prime advantages: protective atmosphere brazing (where high purity gases or vacuum negate the need for flux) and the ability to control accurately every stage of the heating and cooling cycles. Heating is either through elements, or by gas firing.
- **Dip Brazing**
  This involves immersion of the entire assembly into bath of molten braze alloy or molten flux. In both cases the bath temperature is below the solidification point of the parent metal, but above the melting point of the filler metal.

Advantage as compared to welding:

The important brazing techniques advantages as compared to welding process,

- Heat distortion is very less.
- It can be welded dissimilar metals
- If used for metal with poor weldability.
- It not affects the heat treatment of original materials.

Brazing applications:

- It is used to fastening of radiators, electrical parts, tanks, pipe fitting, axles and more.
- it can be weld to dissimilar metal, cast metal and ferrous metal.
- it is used to joint of band saws in bicycle parts such as rims and frames.

TABLE 1. COMPARISON BETWEEN WELDING, SOLDERING AND BRAZING PROCESSES

Comparison between Welding, Soldering and Brazing

Sl. No.	Welding	Soldering	Brazing
1.	These are the strongest joints used to bear the load. Strength of a welded joint may be more than the strength of base metal.	These are weakest joint out of three. Not meant to bear the load. Use to make electrical contacts generally.	These are stronger than soldering but weaker than welding. These can be used to bear the load upto some extent.
2.	Temperature required is upto 3800°C of welding zone.	Temperature requirement is upto 450°C.	It may go to 600°C in brazing.
3.	Workpiece to be joined need to be heated till their melting point.	No need to heat the workpieces.	Workpieces are heated but below their melting point.

9. HEAT-AFFECTED ZONE (HAZ)

The Heat-Affected Zone (HAZ) refers to a non-melted area of metal that has experienced changes in its material properties as a result of exposure to high temperatures. The alterations in material properties are usually a result of welding or high-heat cutting procedures. The HAZ is identified as the area between the weld or cut and the base metal. These areas can vary in size and severity depending on the properties of the materials involved, the intensity and concentration of heat, and the process employed.

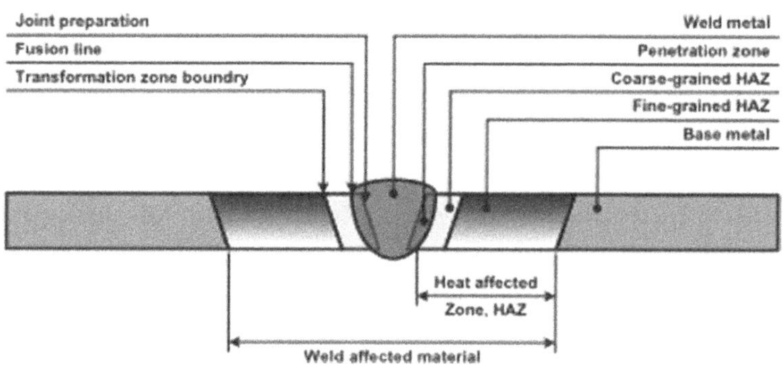

Fig.7. Heat-Affected Zones

What are the Causes of Heat-Affected Zones?

During welding operations, the HAZ may range from small to large depending on the rate of heat input. Welding processes with high rates of heat input (i.e. fast heating) have faster cooler rates compared to welding processes with low rates of heat input (i.e. slow heating) and thus, have smaller HAZs. Conversely, a process with low rates of heat input will result in a larger HAZ. The size of a HAZ also increases as the speed of the welding process decreases. HAZ problems can be mitigated by performing a pre- and/or post-weld heat treatment. Weld geometry also plays a role in the size of the HAZ.

During high-temperature cutting operations, the depth of the HAZ is associated with the cutting process, cutting speed, material properties, and material thickness. Similar to the results of welding processes, cutting processes that operate at high temperatures and slow speeds tend to lead to large HAZs. Furthermore, cutting processes that operate at high speeds tend to reduce the width of the HAZ.

What are the Effects of Heat-Affected Zones?

Because the HAZ experiences sufficient heat for a long enough period of time, the layer undergoes microstructure and property changes that differ from the parent metal. These property changes are usually undesirable and ultimately serve as the weakest part of the component. For example, the micro structural changes can lead to residual stresses, reduced material strength, increased brittleness, and decreased resistance to corrosionand/orcracking. As a result, many failures occur in the HAZ.

HAZ Areas

While the HAZ occurs between the weld or cutting zone and the unaffected parent metal, the HAZ itself has different areas according to how close they are to where the cutting or welding heat was applied to the material.

The cutting or weld zone is the liquid region where the process itself takes place and is adjoined by the fusion boundary. The fusion boundary is the border of the fusion zone where the liquid and solid phases of the metal coexist. Further from the weld or cutting zone is the HAZ itself, which is where the non-melted parent metal has experienced changes to the microstructure. The HAZ can be broken down into the grain growth zone (closest to the heated area), the recrystallized zone, the partially transformed zone and the tempered zone. Outside of these HAZ areas is the unaffected base material.

The various HAZ areas are formed by differing temperatures in the base metal further from the weld or cut itself. This can, for example, be clearly seen by a series of visible colored bands, caused by surface oxidation, near a weld in stainless steel. These different colors, also known as heat tint, offer an approximate indication of the temperature reached by the metal.

10. WELDING DEFECTS

Welding Defects can be defined as the irregularities formed in the given weld metal due to wrong welding process or incorrect welding patterns, etc. The defect may differ from the desired weld bead shape, size, and intended quality. Welding defects may occur either outside or inside the weld metal. Some of the defects may be allowed if the defects are under permissible limits but other defects such as cracks are never accepted.

External Welding Defects:

1. Weld Crack
2. Undercut
3. Spatter
4. Porosity
5. Overlap
6. Crater

Internal Welding Defects:

1. Slag Inclusion
2. Incomplete Fusion
3. Necklace cracking
4. Incompletely filled groove or Incomplete penetration

External Welding Defects

The various types of external defects with their causes and remedies are listed below:

1. Weld Crack

This is the most unwanted defect of all the other welding defects. Welding cracks can be present at the surface, inside of the weld material or at the heat affected zones.

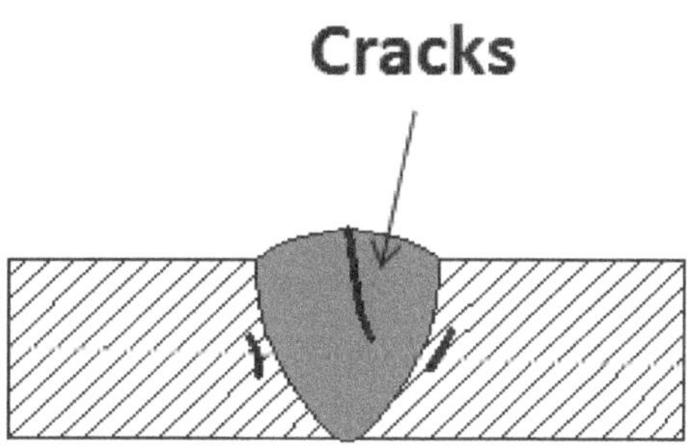

Fig.8. Weld Crack

Causes of Weld Crack:

1. Poor ductility of the given base metal.
2. The presence of residual stress can cause a crack on the weld metal.
3. The rigidity of the joint which makes it difficult to expand or contract the metals.
4. If there is high content on sulfur and carbon then also the cracks may appear.
5. Using hydrogen as a shielding gas while welding ferrous materials.

Remedies for Weld crack:

1. Using appropriate materials may decrease the chances of crack.
2. Preheating the weld and reducing the cooling speed joint helps in reducing crack.
3. Reduce the gap between the weld joints by using reasonable weld joints.

4. While welding releases the clamping force slowly which increases fill to capacity of welding material.

2. Undercut

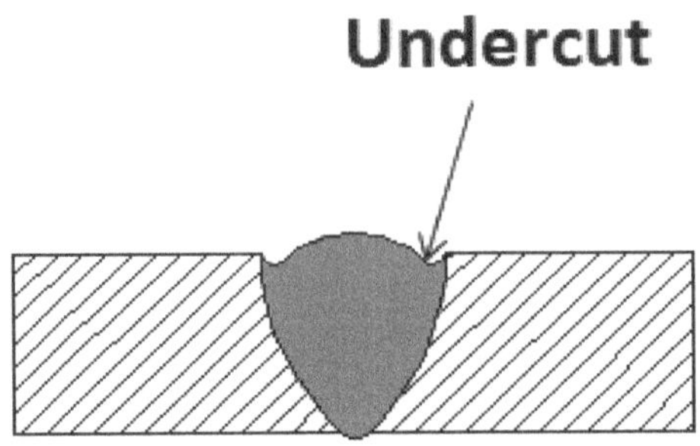

Fig.9. Undercut

When the base of metal melts away from the weld zone, then a groove is formed in the shape of a notch, then this type of defect is known as Undercut. It reduces the fatigue strength of the joint.

Causes of Undercut:

1. If the arc voltage is very high then this defect may occur.
2. If we use the wrong electrode or if the angle of the electrode is wrong, then also the defect may form.
3. Using a large electrode is also not advisable.
4. High electrode speed is also one of the reasons for this defect.

Remedies for Undercut:

1. Reduce the arc length or lower the arc voltage.
2. Keep the electrode angle from 30 to 45 degree with the standing leg.
3. The diameter of the electrode should be small.
4. Reduce the travel speed of the electrode.

3. Spatter

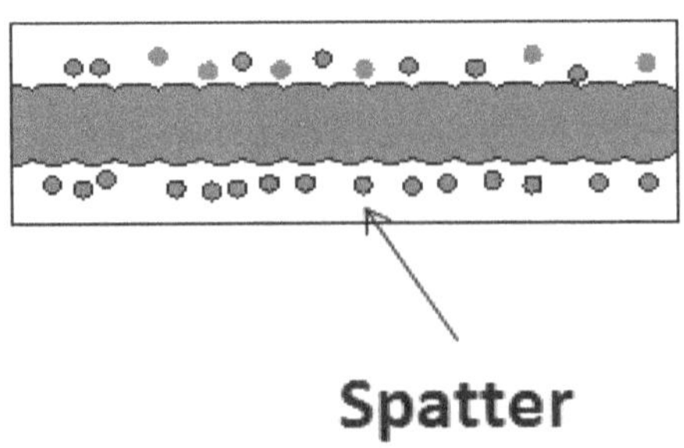

Fig.10. Spatter

When some metal drops are expelled from the weld and remain stuck to the surface, then this defect is known as Spatter.

Causes of Spatter:

1. High Welding current can cause this defect.
2. The longer the arc the more chances of getting this defect.
3. Incorrect polarity.
4. Improper gas shielded may also cause this defect.

Remedies for Spatter:

1. Reducing the arc length and welding current
2. Using the right polarity and according to the conditions of the welding.
3. Increasing the plate angle and using proper gas shielding.

4. Porosity

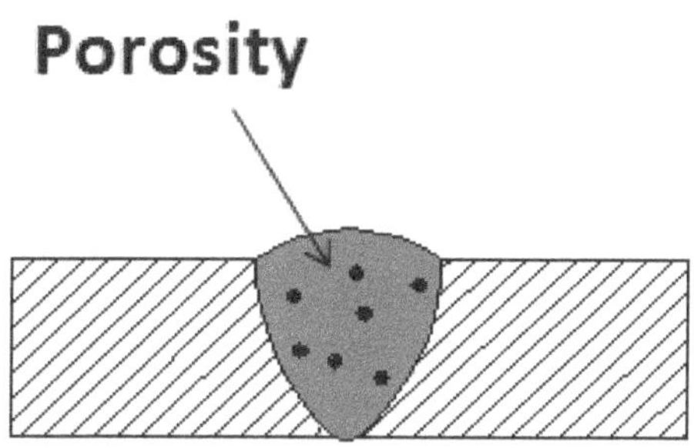

Fig.11. Porosity

Porosity in the condition in which the gas or small bubbles gets trapped in the welded zone.

Causes of Porosity:

1. It occurs when the electrode is not coated properly.
2. Using a longer arc may also increase its chances.
3. Increased welding currents.
4. Rust or oil on the welding surface.

Remedies for porosity:

1. Proper selection of the electrode.
2. Decreasing the welding current.
3. Using smaller arc and slowing the process to allow the gases to escape.
4. Remove rust or oil from the surface and use a proper technique.

5. Overlap

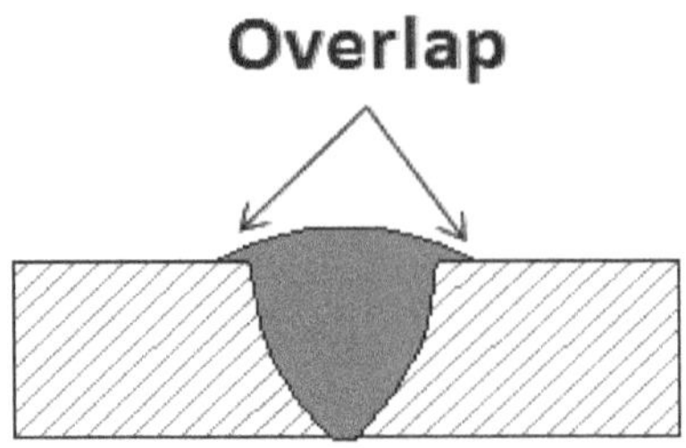

Fig.12. Over lap

When the weld face extends beyond the weld toe, then this defect occurs. In this condition the weld metal rolls and forms an angle less than 90 degrees.

Causes of Overlap:

1. Improper welding technique.
2. By using large electrodes this defect may occur.
3. High welding current

Remedies for Overlap:

1. Using a proper technique for welding.
2. Use small electrode.
3. Less welding current.

6. Crater

It occurs when the crater is not filled before the arc is broken, which causes the outer edges to cool faster than the crater. This causes a stress and then crack is formed.

Causes of the crater:

1. Incorrect torch angle.
2. Use of large electrode:
3. Improper welding technique

Remedies for crater:

1. Using a proper torch angle may reduce the stress on the metal
2. Using a small electrode may also decrease the crater.
3. Use a proper technique.

Internal Welding Defects

The various types of internal welding defects with their causes and remedies are listed below:

1. Slag Inclusion

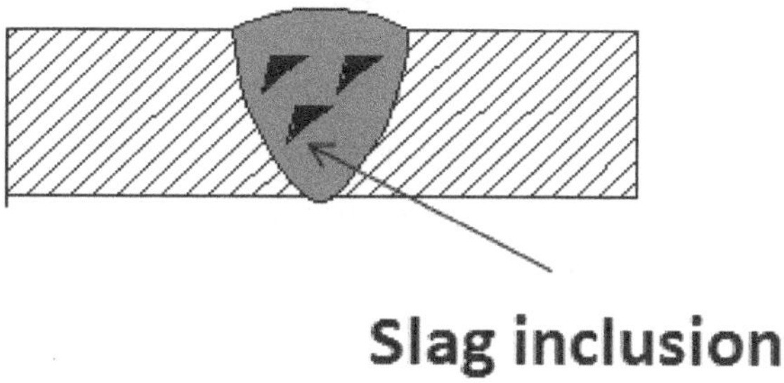

Fig.13. Slag inclusion

If there is any slag in the weld, then it affects the toughness and metal weldability of the given material. This decreases the structural performance of the weld material. Slag is formed on the surface of the weld or between the welding turns.

Causes of Slag:

1. Slag is formed if the welding current density is very small, as it does not provide the required amount of heat for melting the metal surface.
2. If the welding speed is too fast then also slag may occur.
3. If the edge of the weld surface is not cleaned properly then also slag may form.
4. Improper welding angle and travel rate of welding rod.

Remedies for Slag Inclusion:

1. Increase the current density
2. Adjust the welding speed so that the slag and weld pool do not mix with each other.
3. Clean the weld edges and remove the slags of previous weld layers
4. Have a proper electrode angle and travel rate.

2. Incomplete Fusion

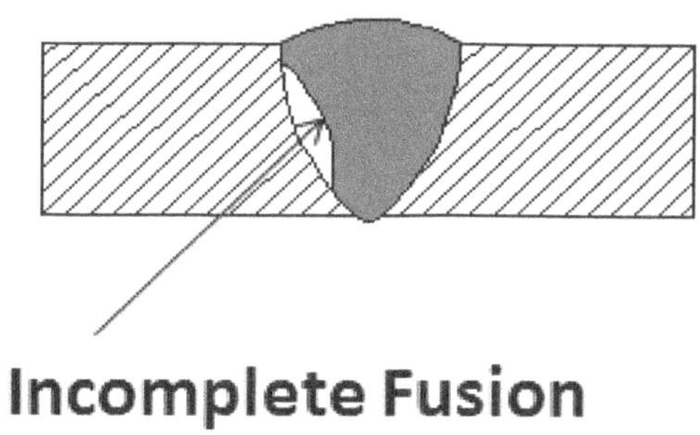

Fig.14. Incomplete fusion

Incomplete fusion occurs when the welder does not accurately weld the material and the metal pre solidifies which leads to a gap which is not filled with the molten metal.

Causes of Incomplete fusion:

1. It occurs because of the low heat input.
2. When the weld pool is very large and runs ahead of the arc.
3. When the angle of the joint is too low.
4. Incorrect electrode and torch angle may also lead to incomplete fusion.
5. Improper bead position.

Remedies for Incomplete Fusion:

1. Increasing the welding current and decreasing the travel speed helps in removing the chances of incomplete fusion.
2. Reducing the deposition rate.
3. Increasing the joint angle.
4. Try to position the electrode and torch angle properly so that the edges of the plate melt away.
5. Positioning the bead properly so that the sharp edges with other beads can be avoided.

3. Necklace Cracking

It occurs in the use of electron beam welding where the weld does not penetrate properly. Therefore, the molten metal does not flow into the cavity and results in a cracking known as "Necklace Cracking".

Causes of Necklace Cracking:

1. Improper welding technique.
2. It occurs in materials such as nickel base alloys, stainless steel, carbon steels and Tin alloys.
3. Using high speed of electron beam welding

Remedies for Necklace Cracking:

1. Using a proper welding technique reduce the chances of necklace cracking.
2. Using proper materials for welding.
3. Using a constant speed during the welding process.
4. Improper welding technique
5. Incompletely Filled Groove or Incomplete Penetration

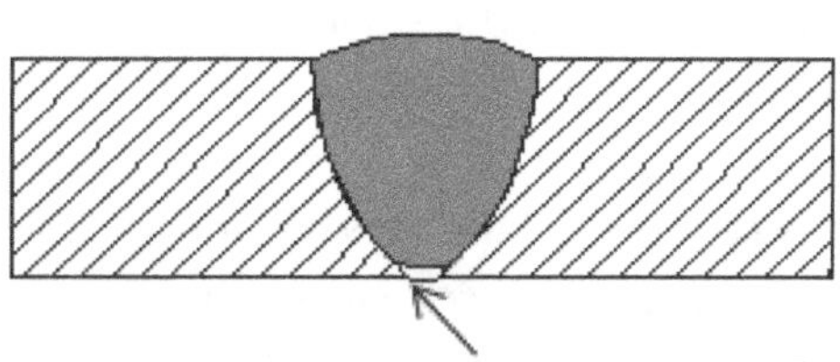

Fig.15. Incomplete penetration

These defects occur only in the butt welds where the groove of the metal is not filled completely. It is also called as incomplete penetration defect.

Causes of an Incomplete filled groove are:

1. Less deposition of the weld metal
2. Use of improper size of the electrode
3. Improper welding technique

Remedies for Incomplete filled groove are:

1. More deposition of the weld metal.
2. Use a proper size of the electrode.
3. By using a proper welding technique.

11. WELDING INSPECTION

Destructive Testing of Welds

Destructive weld testing, as the name suggests, involves the physical destruction of the completed weld in order to evaluate its characteristics. This method of testing is used frequently for a number of applications. Some of these applications include welding procedure qualification and welder performance qualification testing, sampling inspection of production welds, research inspection, and failure analysis work. A number of destructive weld testing methods are used to determine weld integrity or performance. Typically they involve sectioning and/or breaking the welded component and evaluating various mechanical and/or physical characteristics. We shall briefly examine some of the more common methods of this type of welding inspection. We shall consider the macro etch test, the fillet weld break test, the transverse tension test, and the guided bend test. We shall consider how they are used, and what types of weld characteristics they are designed to determine. We shall examine their advantages over other inspection methods and their limitations.

Macro Etch Testing - This method of testing typically involves the removal of small samples of the welded joint. These samples are polished across their cross-section and then etched using some type of mild acid mixture, dependent on the base material used. The acid etch provides a clear visual appearance of the internal structure of the weld. Particular interest is often shown at the fusion line, this being the transition between the weld and the base material. Such items as depth of penetration, lack of fusion, inadequate root penetration, internal porosity, cracking and inclusions can be detected during inspection of the etched sample. This type of inspection is obviously a snapshot of the overall weld length quality when used for sampling inspection of production welds. This type of testing is often used extremely successfully to pinpoint welding problems such as crack initiation, when used for failure analyses.

Fillet Weld Break Test – This type of testing involves breaking a sample fillet weld that is welded on one side only. The sample has load applied to its unwelded side, transverse to the weld and directed to its unwelded side (typically in a press). The load is increased until the weld has failed. The failed sample is then inspected to establish the presence and extent of any welding discontinuities. This test will provide a good indication as to the extent of discontinuities within the entire length of weld tested (normally 6 to 12 inches) rather that a cross-sectional snapspot like the macro etch test. This type of weld inspection can detect such items as lack of fusion, internal porosity and slag inclusions. This testing method is often used in conjunction with the macro etch test. These two testing methods complement each other by providing information on similar characteristics in different detail and in different ways.

Transverse Tension Test – Since a large proportion of design is based on tensile properties of the welded joint, it is important that the tensile properties of the base metal, the weld metal, the bond between the base and the weld, and the heat-affected zone conform to the design requirements. Tensile strength of the welded joint is obtained by pulling specimens to failure. Tensile strength is determined by dividing the maximum load required during testing by the cross-sectional area. The result will be in units of tension per cross-sectional area. This test is nearly always required as part of the mechanical testing when qualifying welding procedure specifications for groove welds.

Guided Bend Test – This is a test method in which a specimen is bent to a specified bend radius. Various types of bend tests are used to evaluate the ductility and soundness of welded joints. Guided bend tests are usually taken transverse to the weld axis and may be bent in plunger type test machines or in wrap-around bend test jigs. Face bend tests are made with the weld face in tension, and root bend tests are made with the weld root in tension. When bend testing thick plates, side bend test specimens are usually cut from the welded joint and bent with the weld cross section in tension. The guided bend test is most commonly used in welding procedure and welder performance qualification tests. This type of testing is particularly good at finding liner fusion defects, which will often open up in the plate surface during the testing procedure.

Non-Destructive Testing Methods

Non-destructive testing (NDT) is a testing and analysis technique used by industry to evaluate the properties of a material, component, structure or system for characteristic differences or welding defects and discontinuities without causing damage to the original part. NDT also known as non-destructive examination (NDE), non-destructive inspection (NDI) and non-destructive evaluation (NDE).

Acoustic Emission Testing (AE)

This is a passive NDT technique, which relies on detecting the short bursts of ultrasound emitted by active cracks under a load. Sensors dispersed over the surface the structure detect the AE. It is even possible to detect AE from plasticization in highly stressed areas before a crack forms. Frequently a method for use during proof tests of a pressure vessel, AE testing is also a continuous Structural Health Monitoring (SHM) method, for example on bridges. Leaks and active corrosion are detectable AE sources too.

Electromagnetic Testing (ET)

This testing method uses an electric current or magnetic field which is passed through a conductive part. There are three types of electromagnetic testing, including eddy current testing, alternating current field measurement (ACFM) and remote field testing (RFT).

Eddy current testing uses an alternating current coil to induce an electromagnetic field into the test piece, alternating current field measurement and remote field testing both use a probe to introduce a magnetic field, with RFT generally used to test pipes.

Ground Penetrating Radar (GPR)

This geophysical NDT method sends radar pulses through the surface of a material or subsurface structure, such as rock, ice, water or soil. The waves are reflected or refracted when they encounter a buried object or material boundary with different electromagnetic properties.

Laser Testing Methods (LM)

Laser testing falls into three categories including holographic testing, laser profilometry and laser shearography.Holographic testing uses a laser to detect changes in the surface of the material which has been subjected to stress such as heat, pressure or vibration. The results are then compared to an undamaged reference sample to show defects.

Laser profilometry uses a high speed rotating laser light source and miniature optics to detect corrosion, pitting, erosion and cracks by detecting changes in the surface via a 3D image generated from the surface topography.Laser shearography uses laser light to create an image before the surface is stressed and a new image is created. These images are compared to one another to determine if any defects are present.

Leak Testing (LT)

Leak testing can be broken down into four different methods - bubble leak testing, pressure change testing, halogen diode testing and mass spectrometer testing.Bubble leak testing uses a tank of liquid, or a soap solution for larger parts, to detect gas (usually air) leaking from the test piece in the form of bubbles.

Only used on closed systems, pressure change testing uses either pressure or a vacuum to monitor the test piece. A loss of pressure or vacuum over a set time span will show that there is a leak in the system. Halogen diode testing also uses pressure to find leaks, except in this case air and a halogen-based tracer gas are mixed together and a halogen diode detection unit (or 'sniffer') is used to locate any leaks.

Mass spectrometer testing uses helium or a helium and air mix inside a test chamber with a 'sniffer' to detect any changes in the air sample, which would indicate a leak. Alternatively, a vacuum can be used, in which case the mass spectrometer will sample the vacuum chamber to detect ionized helium, which will show that there has been a leak.

Magnetic Flux Leakage (MFL)

This method uses a powerful magnet to create magnetic fields which saturate steel structures such as pipelines and storage tanks. A sensor is then used to detect changes in magnetic flux density which show any reduction in material due to pitting, erosion or corrosion.

Microwave Testing

This method is restricted to use on dielectric materials and uses microwave frequencies transmitted and received by a test probe. The test probe detects changes in dielectric properties, such as shrinkage cavities, pores, foreign materials or cracks and displays the results as B or C scans.

Liquid Penetrant Testing (PT)

Liquid penetrant testing involves the application of a fluid with low viscosity to the material to be tested. This fluid seeps into any defects such as cracks or porosity before a developer is applied which allows the penetrant liquid to seep upwards and create a visible indication of the flaw. Liquid penetrant tests can be conducted using solvent removable penetrants, water washable penetrants or post-emulsifiable penetrants.

Magnetic Particle Testing (MT)

This NDT process uses magnetic fields to find discontinuities at or near the surface of ferromagnetic materials. The magnetic field can be created with a permanent magnet or an electromagnet, which requires a current to be applied.

The magnetic field will highlight any discontinuities as the magnetic flux lines produce leakage, which can be seen by using magnetic particles that are drawn into the discontinuity.

Neutron Radiographic Testing (NR)

Neutron radiography uses a beam of low energy neutrons to penetrate into the workpiece. While the beam is transparent in metallic materials most organic materials allow the beam to be seen, allowing the structural and internal components to be viewed and examined to detect flaws.

Radiographic Testing (RT)

Radiographic testing uses radiation passed through a test piece to detect defects. X-rays are commonly used for thin or less dense materials while gamma rays are used for thicker or denser items. The results can be processed using film radiography, computed radiography, computed tomography or digital radiography. Whichever method is used, the radiation will show discontinuities in the material due to the strength of the radiation.

Thermal/Infrared Testing (IRT)

Infrared testing or thermography uses sensors to determine the **wavelength of infrared light emitted from the surface of an object, which can be used to assess its condition.**

Passive thermography uses sensors to measure the wavelength of the emitted radiation and if the emissivity is known or can be estimated, the temperature can be calculated and displayed as a digital reading or as a false colour image. This is useful for detecting overheating bearings, motors or electrical components and is widely used to monitor heat loss from buildings.

Active thermography induces a temperature gradient through a structure. Features within it that affect the heat flow result in surface temperature variations that can be analysed to determine the condition of a component. Often used to detect near surface delaminations or bonding defects in composites.

Ultrasonic Testing (UT)

Ultrasonic Testing entails the transmission of high frequency sound into a material to interact with features within the material that reflect or attenuate it. Ultrasonic testing is broadly divided into Pulse Echo (PE), Through Transmission (TT) and Time of Flight Diffraction (ToFD).

Pulse Echo Inspection

This technique introduces a sound beam into the test material surface. The sound will travel through the part, either reaching the rear wall of the material and then returning to the transducer or returning early when reflected from a discontinuity within the part. If the acoustic velocity is known, the time interval recorded is then used to derive the distance travelled in the material.

Through Transmission Testing

TT uses separate transducers to emit and receive the sound. The transmit probe is positioned one side of the test sample and the receive transducer is positioned on the other side. As the sound passes through the component, it is attenuated by features within it, such as porosity. Thickness measurement is not normally possible with this technique.

Time of Flight Diffraction (ToFD)

Diffraction is the process of a change in wavelength in sound as it interacts with a discontinuity in a material. This mechanism is used in situations where a true reflection cannot be obtained but sufficient diffraction occurs to alter the time of flight of the sound in a pitch-catch arrangement. This method is used to detect the tip of a defect that resides perpendicular the probe contact surface. ToFD is also used for rear wall inspection for detection of corrosion.

Immersion Testing

The requirement to wet couple the ultrasound probe to the part can be a challenge for large or complex geometric samples. For convenience these parts are immersed in water – typically in an immersion tank. This method is usually enhanced by actuators that move the part and/or the probe within the tank during ultrasonic inspection.

Air Coupled Testing

Certain inspections and materials cannot tolerate the application of wet coupled and so in certain circumstances air coupled ultrasound testing may be performed. This entails the application of sound through an air gap. This typically entails the use of lower frequency inspection.

Electromagnetic Acoustic Transducer (EMAT) Testing

EMAT Testing is a type of non-contact inspection method that uses electromagnetic sound generation and reception without immediate contact or wet coupling with the part. EMATs are of particular use for excessively hot, cold, clean, or dry environments. As with conventional ultrasound, EMATs can produce normal and angled beams as well as other modes, such as guided waves.

Guided Wave Testing (GW)

Ideal for testing pipes over long distances, guided wave testing uses ultrasonic wave forms to reflect changes in the pipe wall, which are then sent to a computer for control and analysis. Guided wave testing can be carried out using medium or long range tests - guided wave medium range ultrasonic testing (GW MRUT) and guided wave long range ultrasonic testing (GW LRUT). GW MRUT techniques cover an area of 25mm to 3000mm, while GW LRUT covers distances greater than this and can be used to inspect areas over hundreds of metres from one location.

Advanced Ultrasonic Methods

Automated Inspection

The benefit of automation is achieved by the integration of NDT sensors with standard commercially available industrial robots as well as collaborative robots, also known as "cobots". Custom written software for acquiring and visualizing data creates a seamless and intuitive user experience that can be adapted to specific needs.

TWI has developed several highly capable automated inspection systems suited to both research and development work as well as production inspection.

Phased Array Ultrasonic Testing (PAUT)

PAUT probes are different from conventional UT probes in that they consist of an array of individual elements that can be pulsed independently. By controlling the times at which each element is fired, sound beams can be focused or steered. By sweeping the beam through a range of angles or depths, cross-sectional views can be generated using one probe where several probe and wedge combinations may have been required with conventional UT. A virtual probe can be created from a number of elements and this can be electronically indexed along the length of the array to create a wide paintbrush scan.

Full Matrix Capture (FMC)

FMC is an evolution of the PAUT technique and uses the same probes. Its main advantage is that there is no need to focus or steer the beam as the entire area of interest is in focus. It is also relatively tolerant of misaligned flaws and structural noise. This makes it very easy to set up and use. The disadvantage is that the file sizes are very large and the acquisition speed can be slower than with PAUT.

Virtual Source Aperture (VSA)

VSA is a variation on FMC that retains most of the advantages of its superior image quality, but with greatly reduced file sizes and acquisition speeds that can exceed that of PAUT.

Vibration Analysis (VA)

This process uses sensors to measure the vibration signatures from rotating machinery in order to assess the condition of the equipment. The types of sensors used include displacement sensors, velocity sensors, and accelerometers.

Visual Testing (VT)

Visual testing also known as visual inspection is one of the most common techniques which involves the operator looking at the test piece. This can be aided by the use of optical instruments such as magnifying glasses or computer-assisted systems (known as 'Remote Viewing').

This method allows for the detection of corrosion, misalignment, damage, cracks, and more. Visual testing is inherent in most other types of NDT as they will generally require an operator to look for defects.

UNIT – IV

METALWORKING PROCESSES-I

1. INTRODUCTION OF METAL WORKING

Metalworking is the process of working with metals to create individual parts assemblies, or large-scale structures.

Manufacturing Processes can be classified as

i) Casting
ii) Welding
iii) Machining
iv) Mechanical working
v) Powder Metallurgy
vi) Plastic Technology etc.,

Classification of Metal Working Processes

1. Based on methods

i. Rolling
ii. Forging
iii. Extrusion
iv. Wire Drawing
v. Sheet Metal Forming

2. Based on Temperature of Working

i. Hot Working
ii. Cold Working
iii. Warm Working

3. Based on the applied stress

i. Direct Compressive Stress
ii. Indirect Compressive Stress
iii. Tensile Stress
iv. Bending Stress

2. HOT WORKING AND COLD WORKING OF METALS

HOT WORKING

Mechanical working processes which are done above recrystallisation temperature of the metal are known as hot working processes. Some metals, such as lead and tin, have a low recrystallisation temperature and can be hot-worked even at room temperature, but most commercial metals require some heating. However, this temperature should not be too high to reach the solidus temperature; otherwise the metal will burn and become unsuitable for use. In hot working, the temperature of completion of metal working is important since any extra heat left after working aid in grain growth. This increase in size of the grains occurs by a process of coalescence of adjoining grains and is a function of time and temperature. Grain growth results in poor mechanical properties. If the hot working is completed just above the recrystallisation temperature then the resultant grain size would be fine.

EFFECT OF HOT WORKING ON MECHANICAL PROPERTIES OF METALS

1. This process is generally performed on a metal held at such a temperature that the metal does not work-harden. A few metals e.g., Pb and Sn (since they possess low crystallization temperature) can be hot worked at room temperature.
2. Raising the metal temperature lowers the stresses required to produce deformations and increases the possible amount of deformation before excessive work hardening takes place.
3. Hot working is preferred where large deformations have to be performed that do not have the primary purpose of causing work hardening.
4. Hot working produces the same net results on a metal as cold working and annealing. It does not strain harden the metal.
5. In hot working processes, compositional irregularities are ironed out and nonmetallic impurities are broken up into small, relatively harmless fragments, which are uniformly dispersed throughout the metal instead of being concentrated in large stress-raising metal working masses.
6. Hot working such as rolling process refines grain structure. The coarse columnar dendrites of cast metal are refined to smaller equiaxed grains with corresponding improvement in mechanical properties of the component.
7. Surface finish of hot worked metal is not nearly as good as with cold working, because of oxidation and scaling.
8. One has to be very careful as regards the temperatures at which to start hot work and at which to stop because this affects the properties to be introduced in the hot worked metal.
9. Too high a temperature may cause phase change and overheat the steel whereas too low temperature may result in excessive work hardening.
10. Defects in the metal such as blowholes, internal porosity and cracks get removed or welded up during hot working.
11. During hot working, self-annealing occurs and recrystallisation takes place immediately following plastic deformation. This self-annealing action prevents hardening and loss of ductility.

MERITS OF HOT WORKING

1. As the material is above the recrystallisation temperature, any amount of working can be imparted since there is no strain hardening taking place.

2. At a high temperature, the material would have higher amount of ductility and therefore there is no limit on the amount of hot working that can be done on a material. Even brittle materials can be hot worked.
3. In hot working process, the grain structure of the metal is refined and thus mechanical properties improved.
4. Porosity of the metal is considerably minimized.
5. If process is properly carried out, hot work does not affect tensile strength, hardness, corrosion resistance, etc.
6. Since the shear stress gets reduced at higher temperatures, this process requires much less force to achieve the necessary deformation.
7. It is possible to continuously reform the grains in metal working and if the temperature and rate of working are properly controlled, a very favorable grain size could be achieved giving rise to better mechanical properties.
8. Larger deformation can be accomplished more rapidly as the metal is in plastic state.
9. No residual stresses are introduced in the metal due to hot working.
10. Concentrated impurities, if any in the metal are disintegrated and distributed throughout the metal.
11. Mechanical properties, especially elongation, reduction of area and izod values are improved, but fibre and directional properties are produced.
12. Hot work promotes uniformity of material by facilitating diffusion of alloy constituents and breaks up brittle films of hard constituents or impurity namely cementite in steel.

DEMERITS OF HOT WORKING

1. Due to high temperature in hot working, rapid oxidation or scale formation and surface de-carburization take place on the metal surface leading to poor surface finish and loss of metal.
2. On account of the loss of carbon from the surface of the steel piece being worked the surface layer loses its strength. This is a major disadvantage when the part is put to service.
3. The weakening of the surface layer may give rise to a fatigue crack which may ultimately result in fatigue failure of the component.
4. Some metals cannot be hot worked because of their brittleness at high temperatures.
5. Because of the thermal expansion of metals, the dimensional accuracy in hot working is difficult to achieve.
6. The process involves excessive expenditure on account of high cost of tooling. This however is compensated by the high production rate and better quality of components.
7. Handling and maintaining of hot working setups is difficult and troublesome.

CLASSIFICATION OF HOT WORKING PROCESSES

The classification of hot working processes is given as under.

1. Hot rolling
2. Hot forging
3. Hot extrusion
4. Hot drawing
5. Hot spinning
6. Hot piercing or seamless tubing
7. Tube Forming and
8. Hot forming of welded pipes

Hot Rolling

Rolling is the most rapid method of forming metal into desired shapes by plastic deformation through compressive stresses using two or more than two rolls. It is one of the most widely used of all the metal working processes. The main objective of rolling is to convert larger sections such as ingots into smaller sections which can be used either directly in as rolled state or as stock for working through other processes.

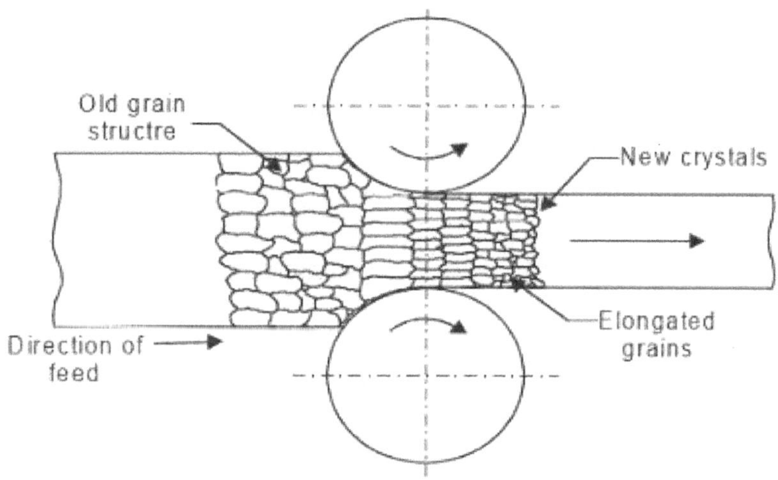

Fig.1. Hot Rolling

Applications of Rolling

In the rail mill, the heavier structural sections and rails are made. Rolling mills produce girders, channels, angle irons and tee-irons. Plate mill rolls slabs into plates. The materials commonly hot rolled are aluminium, copper magnesium, their alloys and many grades of steel.

Hot Piercing or Seamless tubing

Hot piercing is also known as seamless tubing or roll piercing process. The process setup is shown in Figure. It is used for making thin walled round objects. Seamless tube forming is popular and economical process in comparison to machining because it saves material wasted in boring of parts

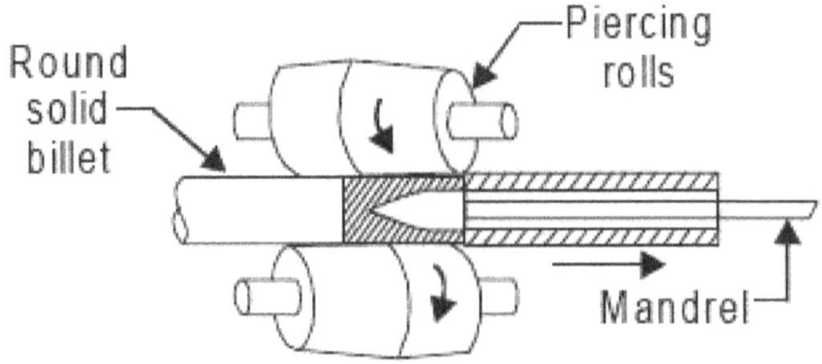

Fig.2. Hot Piercing or Seamless tubing

Hot Extrusion

It is the process of enclosing the heated billet or slug of metal in a closed cavity and then pushing it to flow from only one die opening so that the metal will take the shape of the opening. The pressure is applied either hydraulically or mechanically. Extrusion process is identical to the squeezing of tooth paste out of the tooth paste tube. Tubes, rods, hose, casing, brass cartridge, moulding-trims, structural shapes, aircraft parts, gear profiles, cable sheathing etc. are some typical products of extrusion. Using extrusion process, it is possible to make components, which have a constant cross-section over any length as can be had by the rolling process. The intricacy in parts that can be obtained by extrusion is more than that of rolling, because the die required being very simple and easier to make. Also extrusion is a single pass process unlike rolling. The amount of reduction that is possible in extrusion is large. Generally brittle materials can also be easily extruded. It is possible to produce sharp corners and reentrant angles. It is also possible to get shapes with internal cavities in extrusion by the use of spider dies, which are explained later. The extrusion setup consists of a cylinder container into which the heated billet or slug of metal is loaded. On one end of the container, the die plate with the necessary opening is fixed. From the other end, a plunger or ram compresses the metal billet against the container walls and the die plate, thus forcing it to flow through the die opening, acquiring the shape of the opening. The extruded metal is then carried by the metal handling system as it comes out of the die. The extrusion ratio is defined as the ratio of cross- sectional area of the billet to that of the extruded section. The typical values of the extrusion ratio are 20 to 50. Horizontal hydraulic presses of capacities between 250 to 5500 tonnes are generally used for conventional extrusion. The pressure requirement for extrusion is varying from material to material. The extrusion pressure for a given material depends on the extrusion temperature, the reduction in area and the extrusion speed.

Methods of Hot Extrusion

Hot extrusion process is classified as

1. Direct or forward hot extrusion
2. Indirect or backward hot extrusion
3. Tube extrusion

Direct or Forward Hot Extrusion

Figure (a) shows the direct extrusion operational setup. In this method, the heated metal billet is placed in to the die chamber and the pressure is applied through ram. The metal is extruded through die opening in the forward direction, i.e. the same as that of the ram. In forward extrusion, the problem of friction is prevalent because of the relative motion between the heated metal billet and the cylinder walls. To reduce such friction, lubricants are to be commonly used. At lower temperatures, a mixture of oil and graphite is generally used. The problem of lubrication gets compounded at the higher operating temperatures. Molten glass is generally used for extruding steels.

Indirect or Backward Hot Extrusion

Figure (b) shows the indirect extrusion operational setup. In indirect extrusion, the billet remains stationary while the die moves into the billet by the hollow ram (or punch), through which the backward extrusion takes place. Since, there is no friction force between the billet and the container wall, therefore, less force is required by this method. However this process is not widely used because of the difficulty occurred in providing support for the extruded part.

Tube Extrusion

Figure (c and d) shows the tube extrusion operational setup. This process is an extension of direct extrusion process where additional mandrel is needed to restrict flow of metal for production of seamless tubes. Aluminium based toothpaste and medicated tubes are produced using this process.

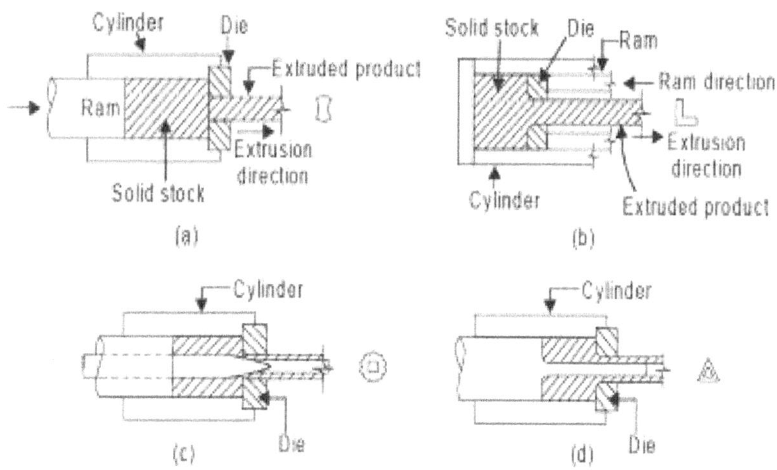

Fig.3. Extrusion process

Hot Drawing

Drawing is pulling of metal through a die or a set of dies for achieving a reduction in a diameter. The material to be drawn is reduced in diameter. Figure 6. is another method used in hot drawing or shaping of materials where the heated blank is placed over the die opening the punch forces the blank through the die opening to form a cup or shell. The multiple dies are also used to accomplish the stages in drawing process. Kitchen utensils and components of food processing industries are manufactured by this process.

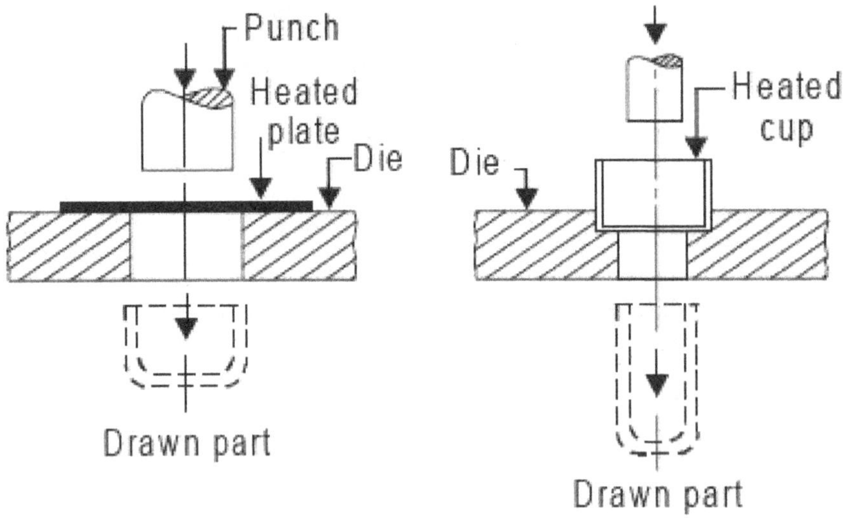

Fig.4. Hot Drawing

Hot Spinning

Hot spinning is a process in which pressure and plastic flow is used to shape material. Spinning may be either hot or cold and is generally carried over a spinning lathe. In both cases, the metal is forced to flow over a rotating shape by pressure of a blunt tool as shown in Figure 7. The amount of pressure of the blunt tool against the disc controls the generated heat, which helps in forming processes.

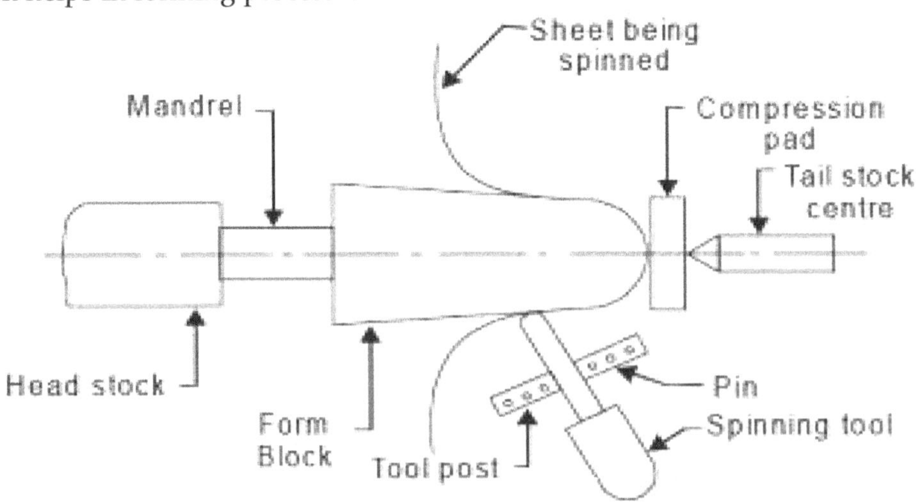

Fig.5. Hot Spinning

Cold Working Processes

Cold working is the plastic deformation of metals below the recrystallisation temperature. In most cases, such cold forming is done at room temperature.The major cold-working operations can be classified basically as squeezing, bending, shearing and drawing.

Cold working is the plastic deformation of metals below the recrystallisation temperature. In most cases of manufacturing, such cold forming is done at room temperature. Sometimes, however, the working may be done at elevated temperatures that will provide increased ductility and reduced strength, but will be below the recrystallisation temperature.

When compared to hot working, cold-working processes have certain distinct advantages:

- No heating required
- Better surface finish obtained
- Superior dimension control
- Better reproducibility and interchangeability of parts
- Improved strength properties
- Directional properties can be minimized

Some disadvantages associated with cold-working processes include:

- Higher forces required for deformation
- Heavier and more powerful equipment required
- Less ductility available
- Metal surfaces must be clean and scale-free
- Strain hardening occurs (may require intermediate anneals)
- Imparted directional properties may be detrimental
- May produce undesirable residual stresses

Table 1: The major cold-working operations

Squeezing	*Bending*	*Shearing*	*Drawing*
Rolling	Angle	Shearing	Bar and tube drawing
Swaging	Roll	Slitting	Wire drawing
Cold forging	Roll forming	Blanking	Spinning
Sizing	Drawing	Piercing	Embossing
Extrusion	Seaming	Lancing	Stretch forming
Riveting	Flanging	Perforating	Shell drawing
Staking	Straightening	Notching	Ironing
Coining		Nibbling	High-energy rate forming
Peening		Shaving	
Burnishing		Trimming	
Die hobbing		Cutoff	
Thread rolling		Dinking	

The major cold-working operations can be classified basically under the headings of squeezing, bending, shearing and drawing, as follows.

Squeezing Processes

Most of the cold-working squeezing processes have identical hot-working counterparts or are extensions of them. The primary reasons for deforming cold rather than hot are to obtain better dimensional accuracy and surface finish. In many cases, the equipment is basically the same, except that it must be more powerful.

Cold Rolling

Cold rolling accounts for by far the greatest tonnage of cold-worked products. Sheets, strip, bars and rods are cold rolled to obtain products that have smooth surfaces and accurate dimensions.

Swaging

Swaging basically is a process for reducing the diameter, tapering, or pointing round bars or tubes by external hammering. A useful extension of the process involves the formation of internal cavities. A shaped mandrel is inserted inside a tube and the tube is than collapsed around it by swaging.

Cold Forging

Extremely large quantities of products are made by cold forging, in which the metal is squeezed into a dive cavity that imparts the desired shape. Cold heading is used for making enlarged sections on the ends of rod or wire, such as the heads on bolts, nails, rivets and other fasteners.

Sizing

Sizing involves squeezing areas of forgings or ductile castings to a desired thickness. It is used principally on basses and flats, with only enough deformation to bring the region to a desired dimension.

Table 2: Difference between hot and cold working processse

Difference between Hot Working and Cold Working:

Hot Working	Cold Working
1. The metal is subjected to mechanical working above RCT	1. The metal is subjected to mechanical working below RCT.
3.Properties are uniform in any direction.	3.The properties are higher in a particular direction than the other.
4. The components will have Isotropic properties.	4. The component will be anisotropic in nature.
5. Energy required for deformation is less.	5. Energy required for deformation is more.
6. No strain hardening takes place in the metal.	6. Strain hardening takes place in the metal.
7. Large components can be hot worked easily.	7. Only small component s can be cold worked
8. Surface oxidation occurs, scaling will be present and surface finish is poor.	8. Surface oxidation is less, scale formation is also less and surface finish is good.
10. Capacity of the equipment is less.	10. The capacity of the equipment required for cold working is less.
11. Needs extra equipment for heating of the metal.	11. Does not need extra equipment as in hot working.
12. Handling of hot metal is difficult.	12. Handling of metal is not very difficult.

3. STRAIN HARDENING

Strain hardening (also called work-hardening or cold-working) is the process of making a metal harder and stronger through plastic deformation. When a metal is plastically deformed, dislocations move and additional dislocations are generated. The more dislocations within a material, the more they will interact and become pinned or tangled. This will result in a decrease in the mobility of the dislocations and a strengthening of the material. This type of strengthening is commonly called cold-working. It is called cold-working because the plastic deformation must occurs at a temperature low enough that atoms cannot rearrange themselves. When a metal is worked at higher temperatures (hot-working) the dislocations can rearrange and little strengthening is achieved.

Strain hardening can be easily demonstrated with piece of wire or a paper clip. Bend a straight section back and forth several times. Notice that it is more difficult to bend the metal at the same place. In the strain hardened area dislocations have formed and become tangled, increasing the strength of the material. Continued bending will eventually cause the wire to break at the bend due to fatigue cracking.

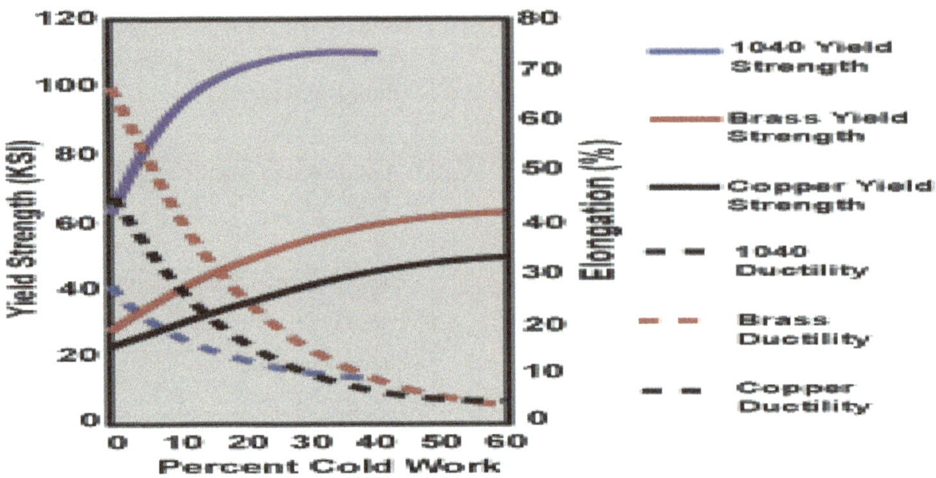

Fig.6. Strain hardening

4. EFFECTS OF ELEVATED TEMPERATURE ON STRAIN HARDENED MATERIALS

When strain hardened materials are exposed to elevated temperatures, the strengthening that resulted from the plastic deformation can be lost. This can be a bad thing if the strengthening is needed to support a load. However, strengthening due to strain hardening is not always desirable, especially if the material is being heavily formed since ductility will be lowered.

Heat treatment can be used to remove the effects of strain hardening. Three things can occur during heat treatment:

- Recovery
- Recrystallization
- Grain growth

Recovery

When a stain hardened material is held at an elevated temperature an increase in atomic diffusion occurs that relieves some of the internal strain energy. Remember that atoms are not fixed in position but can move around when they have enough energy to break their bonds. Diffusion increases rapidly with rising temperature and this allows atoms in severely strained regions to move to unstrained positions. In other words, atoms are freer to move around and recover a normal position in the lattice structure. This is known as the recovery phase and it results in an adjustment of strain on a microscopic scale. Internal residual stresses are lowered due to a reduction in the dislocation density and a movement of dislocation to lower-energy positions. The tangles of dislocations condense into sharp two-dimensional boundaries and the dislocation density within these areas decrease. These areas are called subgrains. There is no appreciable reduction in the strength and hardness of the material but corrosion resistance often improves.

Recrystallization

At a higher temperature, new, strain-free grains nucleate and grow inside the old distorted grains and at the grain boundaries. These new grains grow to replace the deformed grains produced by the strain hardening. With recrystallisation, the mechanical properties return to their original weaker and more ductile states. Recrystallization depends on the temperature, the amount of time at this temperature and also the amount of strain hardening that the material experienced. The more strain hardening, the lower the temperature will be at which recrystallisation occurs. Also, a minimum amount (typically 2-20%) of cold work is necessary for any amount of recrystallisation to occur. The size the new grains is also partially dependant on the amount of strain hardening. The greater the stain hardening, the more nuclei for the new grains, and the resulting grain size will be smaller (at least initially).

Grain Growth

If a specimen is left at the high temperature beyond the time needed for complete recrystallisation, the grains begin to grow in size. This occurs because diffusion occurs across the grain boundaries and larger grains have less grain boundary surface area per unit of volume. Therefore, the larger grains lose fewer atoms and grow at the expense of the smaller grains. Larger grains will reduce the strength and toughness of the material.

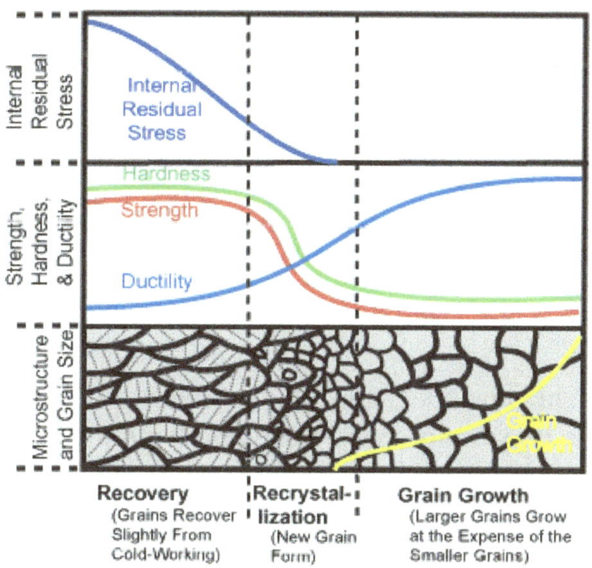

Fig.7. Recovery, recrystallisation and grain growth

5. ROLLING (METALWORKING)

In metalworking, rolling is a metal forming process in which metal stock is passed through one or more pairs of rolls to reduce the thickness and to make the thickness uniform. The concept is similar to the rolling of dough. Rolling is classified according to the temperature of the metal rolled. If the temperature of the metal is above its recrystallization temperature, then the process is known as hot rolling. If the temperature of the metal is below its recrystallization temperature, the process is known as cold rolling. In terms of usage, hot rolling processes more tonnage than any other manufacturing process, and cold rolling processes the most tonnage out of all cold working processes.[1][2] Roll stands holding pairs of rolls are grouped together into rolling mills that can quickly process metal, typically steel, into products such as structural steel (I-beams, angle stock, channel stock), bar stock, and rails. Most steel mills have rolling mill divisions that convert the semi-finished casting products into finished products.

Meaning of Rolling:

The process of shaping metals into semi-finished or finished forms by passing between rollers is called rolling. Rolling is the most widely used metal forming process. It is employed to convert metal ingots to simple stock members like blooms, billets, slabs, sheets, plates, strips etc.

In rolling, the metal is plastically deformed by passing it between rollers rotating in opposite direction. The main objective of rolling is to decrease the thickness of the metal. Ordinarily, there is negligible increase in width, so that the decrease in thickness results in an increase in length.

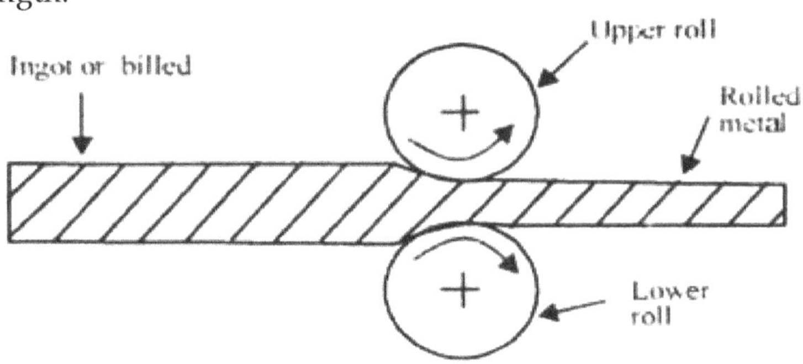

Fig.8. Rolling process

Rolling is done both hot and cold. It is accomplishes in rolling mills. A rolling mill is a complex machine having two or more working rollers, supporting rollers, roll stands, drive motor, reducing gear, flywheel, coupling gear etc.

Rollers may be plain or grooved depends upon the shape of rolled product. The metal changes its shape gradually during the period in which it is in contact with the two rollers.

The range of products that can be produced by rolling is very large. Rolling is a more economical method of deformation than forging when metal is required in long lengths of uniform cross-section.

It is one of the most widely used among all the metal working processes, because of its higher productivity and lower cost. The materials commonly rolled are steel, copper, magnesium, aluminum and their alloys.

Principles of Rolling:

The rolling is a process which consists of passing the metal through a gap between rollers rotating in opposite direction. This gap is smaller than the thickness of the part being worked. Therefore, the rollers compress the metal while simultaneously shifting it forward because of the friction at the roller-metal interfaces.

When the work piece completely passes through the gap between the rollers, it is considered fully worked. As a result, the thickness of the work is decreases while its length and width increases.

However, the increase in width is insignificant and is usually neglected. The Fig. 2.4 shows the simple rolling operation of a plate. The decrease in thickness is called draft, whereas the increase in length is termed as absolute elongation. The increase in width is known as absolute spread.

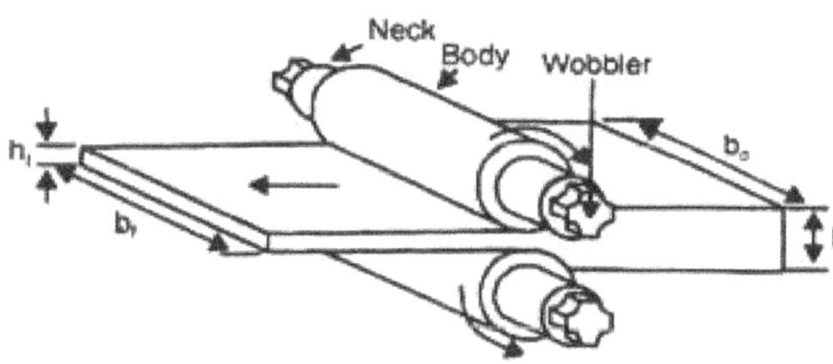

Fig.9. Principles of Rolling

Load and Power Requirement for Rolling:

The deformation zone, state of stress and angle of contact in rolling is shown in Fig. 2.4, (simple rolling of a plate). The main stress system produced in the deformation zone is tri-axial compression. The maximum or principal stress is acting normal to the direction of rolling.

The deformed metal is exerting an equal and opposite force on each of the rolls to satisfy the equilibrium conditions.

Therefore, this force normal to the direction of rolling is important factor considered for designing of the rolls and mill body. This force (F) is also important in determining the power consumption in a rolling process.

Unfortunately, the exact determination of the rolling load and power consumption is a typical task and requires a well knowledge of theory of plasticity and calculus.

Nevertheless, a first approximation of the rolling load can be given by the following equation:

$$F = \bar{y} \times b \times \sqrt{R \times \Delta h} \quad ...(1)$$

where, $\bar{y} = 1.15\,y$ = average yield stress.

and, y = mean yield stress.

Therefore, equation (1) becomes

$$F = 1.15\,y \times b \times \sqrt{R \times \Delta h} \quad ...(2)$$

This equation (2) neglects the friction at the roller-work interface, and therefore gives lower estimation of the rolling load.

Based on the experiments, a multiplication factor of 1.2 is used in modified equation, to consider frictions are:

$$F = 1.2 \times 1.15\,y \times b \times \sqrt{R \times \Delta h} \quad ...(3)$$

Also, the power consumption in rolling process cannot be obtained easily; however, a rough estimation, (in low friction consideration) is given by:

$$\text{h. p.} = \frac{\bar{y} \times b \times R \times \Delta h \times \omega}{550}$$

where, ω = angular velocity of rollers (rad/sec).

$\bar{y}$ = average yield stress (English unit)

R = radius of roller. (English unit)

$\Delta h = h_0 - h_f$ = change in thickness. (English unit)

The various methods of reducing the separating force (F) are following:

(a) Smaller roll diameter (which reduces contact area).
(b) Lower friction.
(c) Higher work piece temperature.
(d) Take 'angle of bite' small (thereby reducing contact area).

6. DEFECTS IN ROLLED PRODUCTS:

A number of defects in the rolled products arise during rolling process. A particular defect is usually arrived with a particular process and does not arise in other processes.

Some of the common defects in rolled products are given below:

(i) Edge Cracking:

Edge cracking generally occurs in rolled ingots, slabs, or plates. This is due to, either limited ductility of the work metal or uneven deformation, especially at the edges.

(ii) Folds:

Folds are a defect generally occurs in plate rolling. This is caused if the reduction per pass is too small.

(iii) Alligatoring:

Alligatoring is the defect, usually occurs in the rolling of slabs (particularly aluminum and alloys). In this defect, the work piece splits along a horizontal plane on exit, with the top and bottom. This defect always occurs when the ratio of slab thickness to the length of contact fall within the range of 1.4 to 1.65. Fig. 2.15. Shows the defect of Alligatoring.

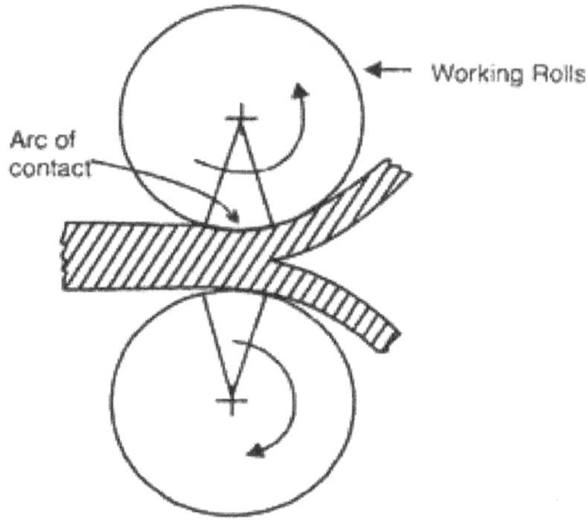

Fig.10. Alligatoring

(iv) Scale Formation:

When the metal is hot rolled, its surface is not smooth and it has scale (oxide) formed over it.

7. TYPES OF ROLLING MILLS

Rolling mills may be classified according to the number and arrangement of the rolls.

a) Two high rolling mills
b) Three high rolling mills
c) Four high rolling mills
d) Tandem rolling mills
e) Cluster rolling mills

Twohigh rolling mills

Two high rolling mills may further classified as

- Reversing mill
- Non reversing mill

A two high rolling mill has two rolls only.

Two high reversing mill:

In two high reversing rolling mills the rolls rotate ist in one direction and then in the other, so that rolled metal may pass back and forth through the rolls several times. This type is used in pluming and slabing mills and for roughing work in plate , rail , structural and other mills.

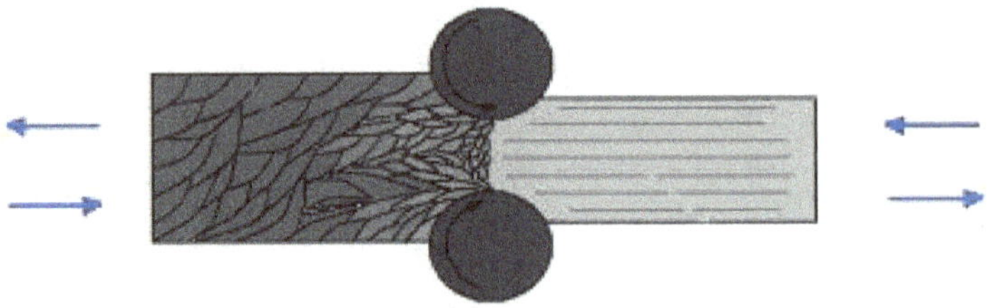

Fig.11.Two high reversing rolling mill

These are more expensive compared to the non reversing rolling mills. Because of the reversible drive needed.

Two high non reversing mill:

In two high non reversing mills as two rolls which revolve continuously in same direction therefore smaller and less costly motive power can be used. However every time material is to be carried back over the top of the mill for again passing in through the rolls. Such an arrangement is used in mills through which the bar passes once and in open train plate mill.

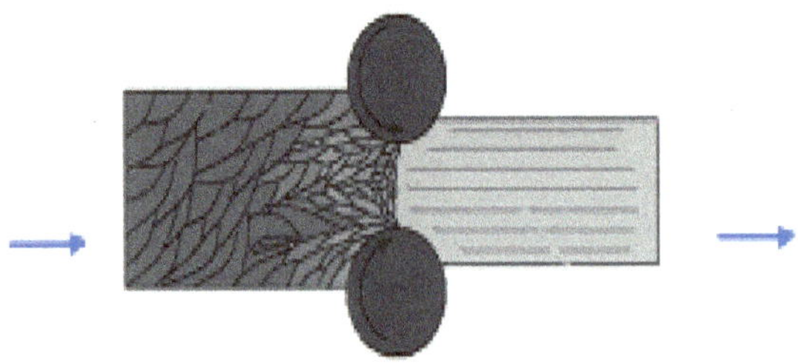

Fig.12.Two high non reversing rolling mill

2: Three high rolling mill:

It consists of a roll stand with three parallel rolls one above the other. Adjacent rolls rotates in opposite direction. So that the material may be passed between the top and the middle roll in one direction and the bottom and middle rolls in opposite one.

In three high rolling mills the work piece is rolled on both the forward and return passes. First of all the work piece passes through the bottom and middle rolls and the returning between the middle and the top rolls.

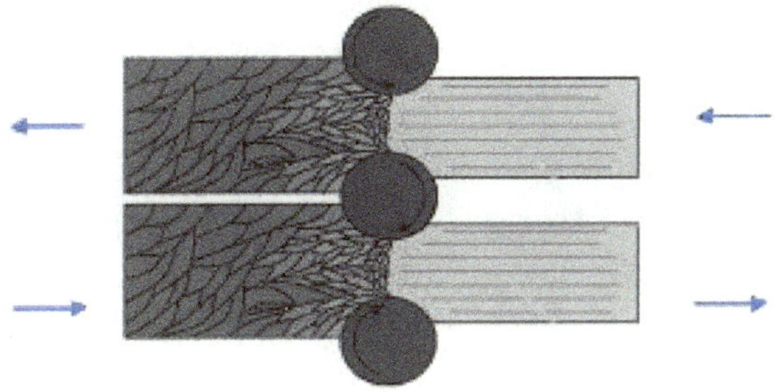

Fig.13.Three high rolling mill

So that thickness is reduced at each pass. Mechanically operated lifted tables are used which move vertically or either side of the stand. So that the work piece fed automatically into the roll gap.

Since the rolls run in one direction only a much less powerful motor and transmission system is required. The rolls of a three high rolling mills may be either plain or grooved to produce plate or sections respectively.

3: Four high rolling mill:

It has a roll stand with four parallel rolls one above the other. The top and the bottom rolls rotate in opposite direction as do the two middle rolls. The two middle are smaller in size than the top and bottom rolls which are called backup rolls for providing the necessary rigidity to the smaller rolls

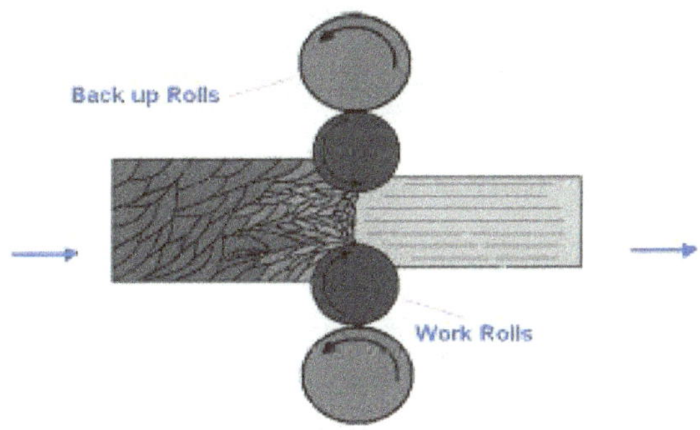

Fig.14. Four high rolling mill

A four high rolling mill is used for the hot rolling of armor and other plates as well as cold rolling of plates, sheets and strips.

4: Tandem rolling mills:

It is a set of two or three stands of roll set in parallel alignment. So that a continuous pass may be made through each one successively with change the direction of material.

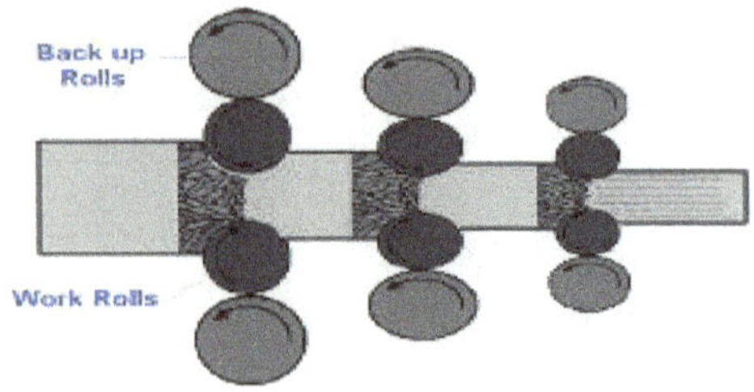

Fig.15. Tandem rolling mill

5: Cluster rolling mills:

It is a special type of four high rolling mill in which each of the two working rolls is backup by two or more of the larger backup rolls for rolling hard in materials. It may be necessary to employ work rolls of a very small diameter but of considerable length. In such cases adequate of the working rolls can be obtained by using a cluster mill.

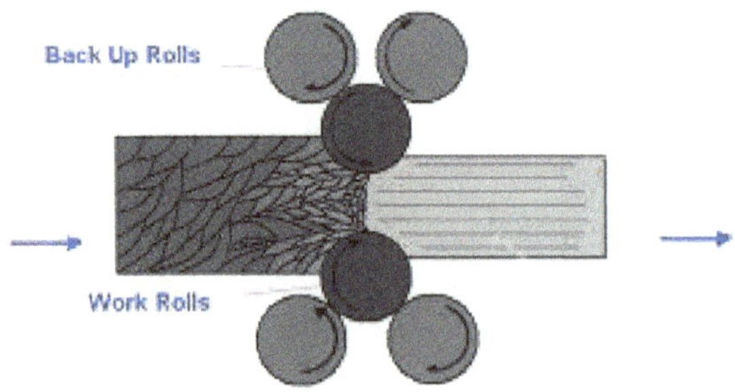

Fig.16. Cluster rolling mills

Application:

- Rolling is used for making hollow seamless tubes, rods etc.
- Large length cross sections are produced by rolling process.
- It is use for mass production of threaded parts like screw, bolts etc.
- Gears can be cut on gear blank by rolling process.
- Construction material, roofing panels, partition **beams**, railroads, etc. are rolling product.
- It is used in automotive industries for manufacture various parts.
- Rings of turbines, bearing and other machines are rolling product.
- Steel sheets, plates are made by rolling process.

Types of Rolling:

Rolling can be classified into following ways.

Thread or Gear Rolling:

This process is used to cut **threads** or **gear** on cylindrical blank. In this process threaded dies are fitted on cylindrical rollers of rolling machine. This threaded roller pressed against cylindrical blank and roll against the faces which displace the material and form threads on cylindrical blank. This process gives high strength and surface finish and used to mass production of screw, bolts etc.

Shape Rolling:

Shape rolling is used to cut various shapes on metal work piece. It does not involve appreciable change in thickness. It is used to rolled I section, T section, railroads etc.

Ring Rolling:

Ring rolling is an another method of rolling which is used to rolled rings of large **bearings, turbines** etc. In this process, two rollers (Main Roller and Idler Roller) are arranged as shown in figure and rotate in same direction to each other. The ring rotates due to rotation of roller and the rollers start move closer to each other. This will increase the diameter of rings. A pair of edge rollers is used to maintain the height of ring which does not allow to metal flow in the direction of height. This process gives high accuracy and material finish.

Tube Piercing:

It is another rolling process in which a stationary mandrel is used at the center of tube and cavity form due to tensile stress in a cylindrical rod when subjected to external compressive stress. In this process two opposite rolls are rotate in opposite direction which compress the tube and feed it against mandrel which create hollow cavity in it. This process is used to making seamless hollow tubes of thick wall.

Skew Rolling:

Skew rolling is used to make balls of ball bearing. In this process round bar or wire is fed directly into special designed rollers which continuously form spherical balls by rolling action. This process is used for mass production of small size spherical balls.

Transverse Rolling:

Transverse rolling is another type of rolling which is used to make tapper shaft, **leaf springs** etc. in this process both rollers are rotate in same direction and the bar is fed transversely. The rolls are fitted with special kind of tapped section which used to form tapped cavity of cylindrical work piece.

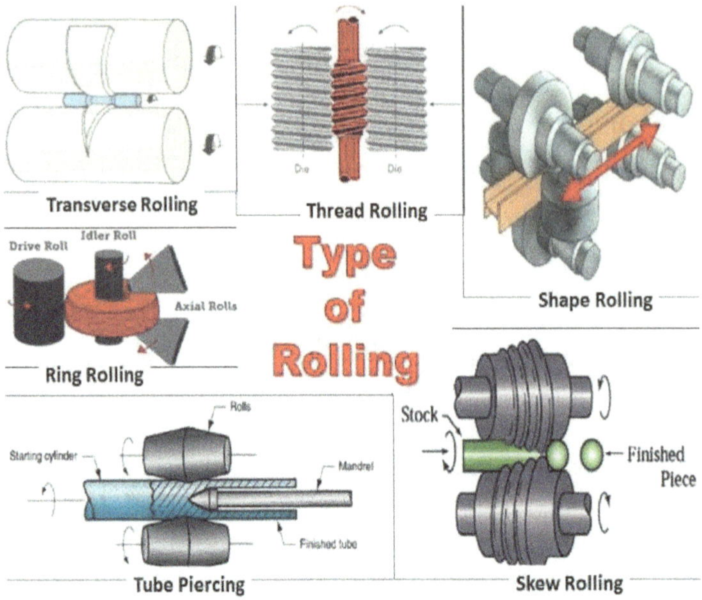

Fig.17. Transverse rolling

Metal Stamping

Metal stamping is a cold-forming process that makes use of dies and stamping presses to transform sheet metal into different shapes. Pieces of flat sheet metal, typically referred to as blanks, is fed into a sheet metal stamping press that uses a tool and die surface to form the metal into a new shape. Production facilities and metal fabricators offering stamping services will place the material to be stamped between die sections, where the use of pressure will shape and shear the material into the desired final shape for the product or component.

This article describes the metal stamping process and steps, presents the types of stamping presses typically employed, looks at the advantages of stamping compared to other fabrication processes, and explains the different types of stamping operations and their applications.

Basic Concepts of Metal Stamping

Metal stamping, also referred to as pressing, is a low-cost high-speed manufacturing process that can produce a high volume of identical metal components. Stamping operations are suitable for both short or long production runs, and be conducted with other metal forming operations, and may consist of one or more of a series of more specific processes or techniques, such as:

- Punching
- Blanking
- Embossing
- Coining

- Bending
- Flanging

Punching and blanking refer to the use of a die to cut the material into specific forms. In punching operations, a scrap piece of material is removed as the punch enters the die, effectively leaving a hole in the workpiece. Blanking, on the other hand, removes a workpiece from the primary material, making that removed component the desired workpiece or blank.

Embossing is a process for creating either a raised or recessed design in sheet metal, by pressing the raw blank against a die that contains the desired shape, or by passing the material blank through a roller die.

Coining is a bending technique wherein the workpiece is stamped while placed between a die and the punch or press. This action causes the punch tip to penetrate the metal and results in accurate, repeatable bends. The deep penetration also relieves internal stresses in the metal workpiece, resulting in no spring back effects.

Bending refers to the general technique of forming metal into desired shapes such as L, U, or V-shaped profiles. The bending process for metal results in a plastic deformation which stresses above the yield point but below the tensile strength. Bending typically occurs around a single axis.

Flanging is a process of introducing a flare or flange onto a metal workpiece through the use of dies, presses, or specialized flanging machinery.

Types of Stamping Operations

Progressive die stamping

Progressive die stamping uses a sequence of stamping stations. A metal coil is fed into a reciprocating stamping press with progressive stamping dies. The die moves with the press, and when the press moves down the die closes to stamp the metal and form the part. When the press moves up, the metal moves horizontally along to the next station. These movements must be precisely aligned as the part is still connected to the metal strip. The final station separates the newly-fabricated part from the rest of the metal. Progressive die stamping is ideal for long runs, because the dies last a long time without becoming damaged, and the process is highly repeatable. Each step in the process performs a different cut, bend, or punching operation on the metal, thus gradually achieving the desired end-product shape and design. It is also a faster process with a limited amount of wasted scrap.

Transfer Die Stamping

Transfer die stamping is similar to progressive die stamping, but the part is separated from the metal trip early on in the process and is transferred from one stamping station to the next by another mechanical transport system, such as a conveyor belt. This process is usually used on larger parts that may need to be transferred to different presses.

Four-Slide Stamping

Four-slide stamping is also called multi-slide or four-way stamping. This technique is best-suited for crafting complex components that have numerous bends or twists. It uses four sliding tools, instead of one vertical slide, to shape the workpiece through multiple deformations. Two slides, or rams, strike the workpiece horizontally to shape it, and no dies are used. Multi-slide stamping can also have more than four moving slides.

Four-slide stamping is a very versatile type of stamping, as different tools can be attached to each slide. It also has a relatively low cost, and production is fast.

Fine Blanking

Fine blanking, also known as fine-edge blanking, is valuable for providing high accuracy and smooth edges. Usually done on a hydraulic or mechanical press, or by a combination of the two, fine blanking operations consist of three distinct movements:

1. Clamping of the workpiece or work material in place
2. Performance of the blanking operation
3. Ejection of the finished part

Fine blanking presses operate at higher pressures than those used in conventional stamping operations, hence tools and machinery need to be designed with these higher operating pressures in mind.

The edges that are produced from fine blanking avoid fractures as produced with conventional tooling and surface flatness can exceed that available from other stamping methods. Since it is a cold extrusion technique, fine blanking is a single-step process, reducing the overall costs of fabrication.

Forming, metal forming, is the metalworking process of fashioning metal parts and objects through mechanical deformation; the workpiece is reshaped without adding or removing material, and its mass remains unchanged. Forming operates on the materials science principle of plastic deformation, where the physical shape of a material is permanently deformed.

Characteristics

Metal forming tends to have more uniform characteristics across its subprocesses than its contemporary processes, cutting and joining.

On the industrial scale, forming is characterized by

- Very high loads and stresses required, between 50 and 2500 N/mm^2 (7-360 ksi)
- Large, heavy, and expensive machinery in order to accommodate such high stresses and loads
- Production runs with many parts, to maximize the economy of production and compensate for the expense of the machine tools

Forming processes

Forming processes tend to be categorized by differences in effective stresses. These categories and descriptions are highly simplified, since the stresses operating at a local level in any given process are very complex and may involve many varieties of stresses operating simultaneously, or it may involve stresses which change over the course of the operation.[3]

Compressive forming involves those processes where the primary means of plastic deformation is uni- or multiaxial compressive loading.

- Rolling, where the material is passed through a pair of rollers
- Extrusion, where the material is pushed through an orifice
- Die forming, where the material is stamped by a press around or onto a die
- Forging, where the material is shaped by localized compressive forces
- Indenting, where a tool is pressed into the workpiece

Tensile forming

Tensile forming involves those processes where the primary means of plastic deformation is uni- or multiaxial tensile stress.

- Stretching, where a tensile load is applied along the longitudinal axis of the workpiece
- Expanding, where the circumference of a hollow body is increased by tangential loading
- Recessing, where depressions and holes are formed through tensile loading

Combined tensile and compressive forming

This category of forming processes involves those operations where the primary means of plastic deformation involves both tensile stresses and compressive loads.

- Pulling through a die
- Deep drawing
- Spinning
- Flange forming
- Upset bulging

Bending

Main article: Bending (metalworking)

This category of forming processes involves those operations where the primary means of plastic deformation is a bending load.

Shearing

Main article: Shear forming

This category of forming processes involves those operations where the primary means of plastic deformation is a shearing load.

Difference between Blanking, Punching & Piercing in Sheet Metal

Blanking

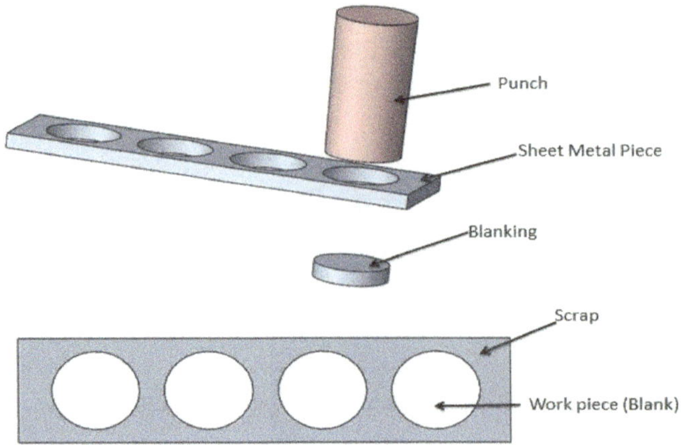

Fig.18. Blanking

Punching blanking and piercing are sheet metal shearing operations to modify existing blank.

The blanking operation produces an enclosed part from a large piece of sheet metal. As a result, the blank part is used as the final product. In blanking , the removed portion is **useful** & the rest of the sheet metal is scrap.

Blanking and piercing are the sheet metal operations which involves punching a hole or removing the blank in the sheet metal using punch and die. Due to shearing the blank is removed. The main difference is in punching the punch is made to hole size and shear is provided on punch. And die is given clearance. Whereas in blanking operation, the die size is made to blank size and shear is provided on the die. And punch is given clearance.
For example disk or plane washers are manufactured using blanking operation.

Punching

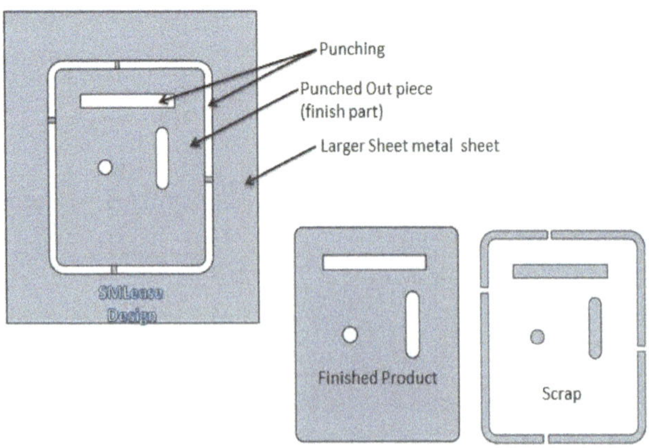

Fig.19.Punching

Punching operation removes scrap from the larger piece of sheet metal. It is a metal removal operation. As a result of punching, a larger piece of sheet metal becomes the final product.

In other words, Combination of punching operations modify existing blank/sheet using metal removing operations.

Piercing

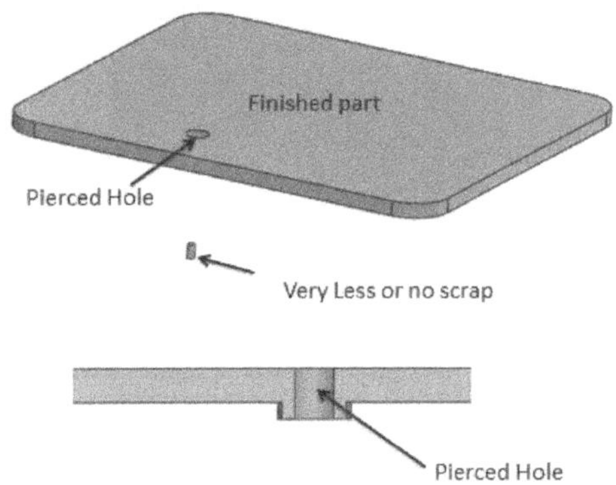

Fig.20.Piercing

Piercing operation produces an extruded hole or slot. As a result of applied force from the piercing tool, sheet metal starts tearing and produces extruded hole/slot.

In other words, Piercing operation produces an extruded hole in sheet metal.Bending is a process by which metal can be deformed by plastically deforming the material and changing its shape. The material is stressed beyond the yield strength but below the ultimate tensile strength. The surface area of the material does not change much. Bending usually refers to deformation about one axis.

Bending Process

Bending is a flexible process by which many different shapes can be produced. Standard die sets are used to produce a wide variety of shapes. The material is placed on the die, and positioned in place with stops and/or gages. It is held in place with hold-downs. The upper part of the press, the ram with the appropriately shaped punch descends and forms the v-shaped bend.

Bending is done using Press Brakes. Press Brakes normally have a capacity of 20 to 200 tons to accommodate stock from 1m to 4.5m (3 feet to 15 feet). Larger and smaller presses are used for specialized applications. Programmable back gages, and multiple die sets available currently can make for a very economical process.

Air Bending is done with the punch touching the workpiece and the workpiece, not bottoming in the lower cavity. This is called air bending. As the punch is released, the workpiece ends up with less bend than that on the punch (greater included angle). This is called spring-back. The amount of spring back depends on the material, thickness, grain and temper. The spring back usually ranges from 5 to 10 degrees. Usually the same angle is used in both the punch and the die to minimize setup time. The inner radius of the bend is the same as the radius on the punch.

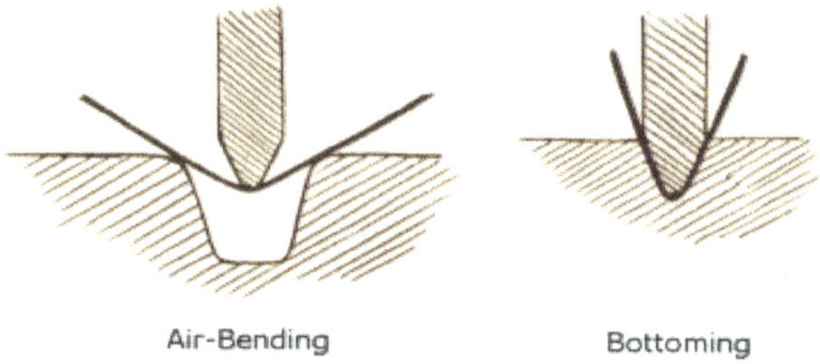

Fig.21 Bending

Bottoming or Coining is the bending process where the punch and the workpiece bottom on the die. This makes for a controlled angle with very little spring back. The tonnage required on this type of press is more than in air bending. The inner radius of the workpiece should be a minimum of 1 material thickness in the case of bottoming; and up to 0.75 material thickness, in the case of coining.

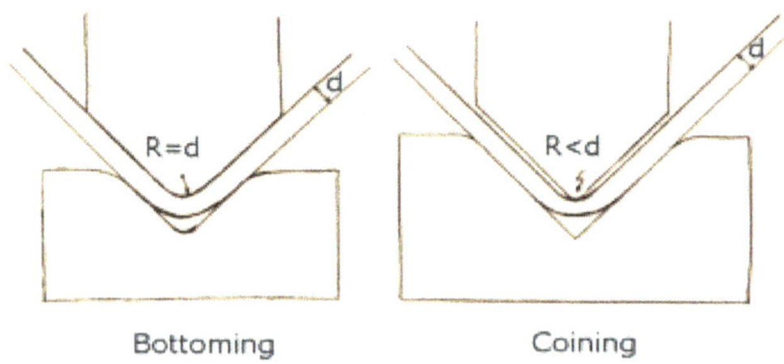

Fig.22 Coining

8. DRAWING PROCESS

Drawing is a metalworking process which uses tensile forces to stretch metal or glass. As the metal is drawn (pulled), it stretches thinner, into a desired shape and thickness. Drawing is classified in two types: sheet metal drawing and wire, bar, and tube drawing. The specific definition for sheet metal drawing is that it involves plastic deformation over a curved axis. For wire, bar, and tube drawing the starting stock is drawn through a die to reduce its diameter and increase its length. Drawing is usually done at room temperature, thus classified a cold working process, however it may be performed at elevated temperatures to hot work large wires, rods or hollow sections in order to reduce forces.

Drawing differs from rolling in that the pressure of drawing is not transmitted through the turning action of the mill but instead depends on force applied locally near the area of compression. This means the amount of possible drawing force is limited by the tensile strength of the material, a fact that is particularly evident when drawing thin wires.

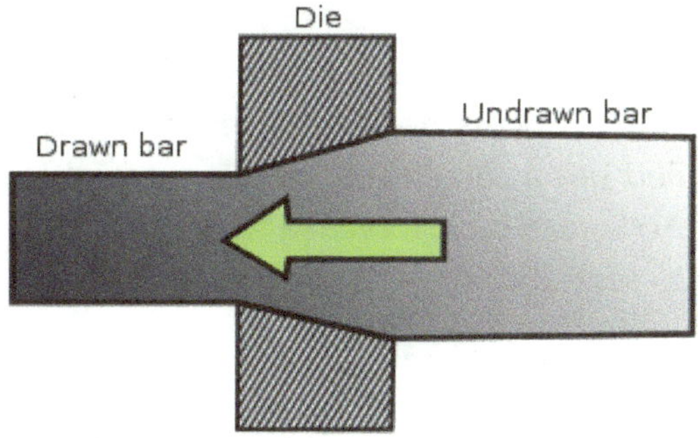

Fig.23Drawing process

Requirement of Drawing Process:

- The material should have sufficient ductility so it can sustain tensile force.
- The material should possess high tensile stress.
- The rod or wire should be properly cleaned and dust or scale free before drawing.
- It should be properly lubricated to reduce friction associated with operation.

TYPES OF DRAWING PROCESS

Wire Drawing:

A wire is a circular, small diameter flexible rod. Wire drawing is acold working process. It is an operation to produce wire of various sizes within certain specific tolerances. This process involves reducing diameter of thick wire by passing it through a series of wire drawing dies with successive die having smaller diameter than the preceding one. Mostly die are made by chilled cast iron, tungsten carbide, diamond or other tool material. The maximum reduction in area of wire is less than 45% in one pass.

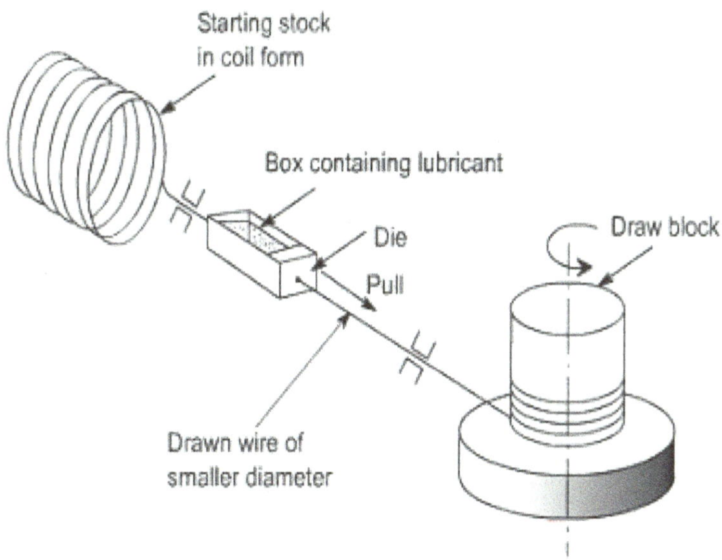

Fig.24 Wire drawing process

Rod Drawing:

Rod drawing is similar process like wire drawing except it is rigid and has larger diameter compare to wire. This process need heavier equipment compare to wire drawing because the wire can be coiled but a rod should be kept straight. The work piece is first fed into die and pulled by a carriage which increase its length and decrease its cross section. Now the rod is to be cut into sections.

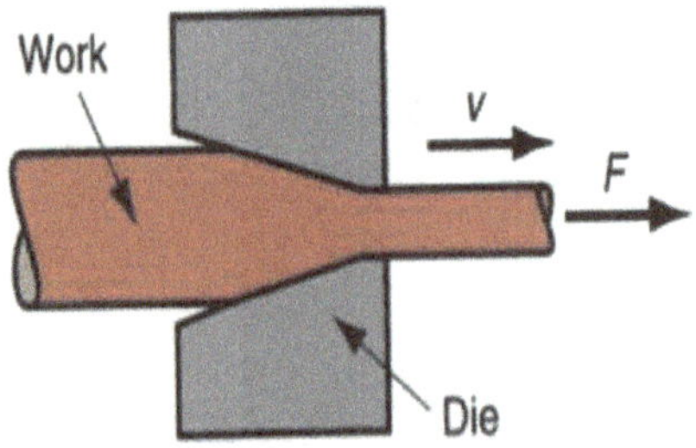

Fig.25 Rod drawing process

Tube Drawing:

Tube drawing is also similar to other two processes except it uses a mandrel to reduce wall thickness and cross section diameter of a tube. This mandrel placed with die and the work piece is pulled by a carriage system as describe in rod drawing. The tube is either circular or rectangular. It also required more than one pass to complete drawing operation.

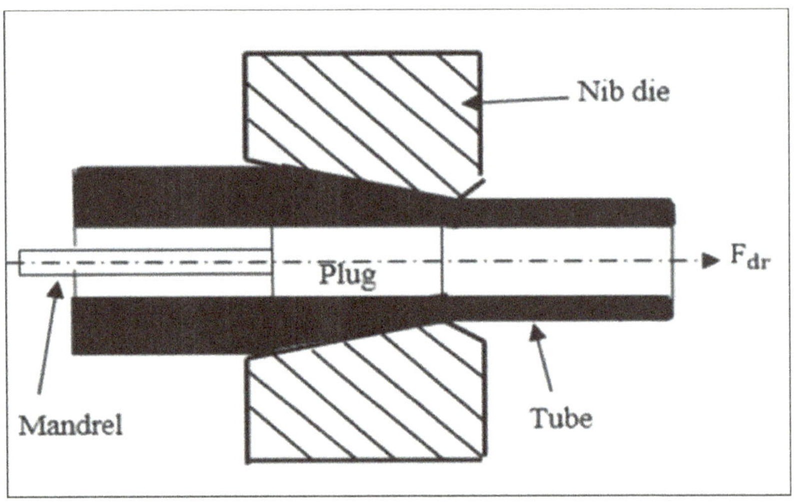

Fig.26 Tube drawing process

Application:

- This process is used for making wire of copper, aluminum etc. which are used in electrical industries.
- Paper clip, **helical spring** etc. are wire drawing product.
- Small diameter rods and tubes are drawing product.
- It is used to produce large length of small cross section.

Meaning of Spinning:

It is a sheet metal forming process in which a metal blank is pressed over a rotating chuck or form mandrel with the help pressing tool to obtain axisymmetric hollow shell.

The sheet metal parts that have circular cross-section can be made by this process. Generally, the shapes produced by spinning can also be manufactured by drawing, compressing or flanging.

However, spinning is usually used for forming large parts that require very large drawing process or when various shapes are needed but only a small number of each shape is required.

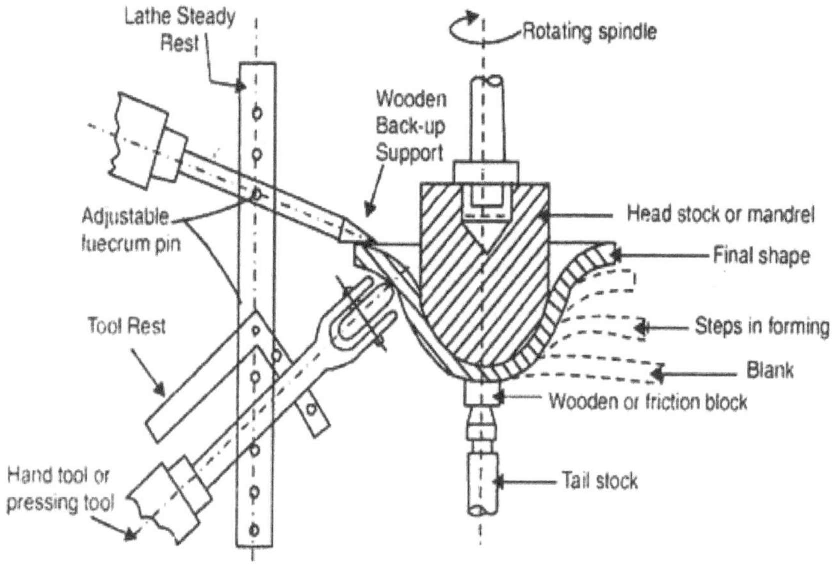

Fig.27 Spinning process

Process of Spinning:

A Schematic of the spinning operation is shown in Fig. 8.1. The set-up essentially consists of a spinning machine similar to a centre lathe

(i) The circular blank is properly centered and pushed by the tail stock against the front of the rotating chuck, usually made of wood. A wooden block is used from the tail-stock to support the blank.

(ii) The chuck is rotated by the head-stock of a spinning lathe machine. The spinning speed depends on the blank material, thickness and complexity of the part required. It is varies from a very slow to a high speed of about 3500 RPM.

(iii) A hard wood or roller type metallic pressing tool is pushed by the operator onto the external surface of the blank.

(iv) Then the tool is moved gradually on the blank so that is conforms to the shape of the form block or mandrel.

(v) The blank slips under the pressing tool, which causes localized deformation. This deformation of the metal is proceeds by a mixture of bending and stretching.

(vi) To avoid wrinkles, a back-up support is applied to the opposite size of the pressing tool.

(vii) Finally, the blank takes the exact shape of the form mandrel.

A modified version of this method involves replacing the operator by a NC (Numerical Controlled) tool. Better surface quality and more uniform thickness are the advantages of this NC spinning over conventional spinning technique.

Types of Spinning Processes:

The Spinning process is classified in two types i.e., hot-spinning and cold-spinning, depending upon whether the blank or work piece has been heated before spinning or not.

(i) Hot Spinning:

In hot spinning, the metal blank is heated to forging temperature and then forming it into the desired shape.

A blunt pressing tool is used which contacts the surface of the rotating part and causes the flow of metal over the form mandrel. This method is generally used for thicker plates and sheets, which do not plastically de-formed at room temperature by pressing tool.

In order to avoid wrinkling at the outer edge a back-up support (hard wood bar) opposite to the tool is used, when working with relatively thin sheets. Hot spinning produces parts such as heads for pressure vessels, refinery equipment's and large tanks.

(ii) Cold Spinning:

Cold spinning process is similar to hot spinning except that the metal blank is worked at room temperature. This method is generally best suited for thin plates and sheets of aluminum and other soft metals.

Cold spinning produces parts such as light reflectors, cooking utensils, liquid containers, radial engine cowling, domestic use hollow parts etc.

Spinning Machines:

Any heavy duty lathe with adequate speed range can be used for spinning process. However, special spinning machines have also been designed fitted with a ball-bearing centre, a tool rest with movable steel pins and a rotating chuck made of hard wood.

Advantages of Spinning:

The spinning process usually used for forming large parts and when various shapes are needed but only a small number of each shape is required.

Some advantages are listed below:

(i) Comparable to Drawing Process:

The spinning process is comparable to drawing process for producing cylindrical, axisymetry parts.

(ii) Economical Process:

The spinning process uses the simple tools; hence it is economical for small lots.

(iii) Best Suited for Large Parts:

Larger parts are much more easily made in spinning than by drawing.

(iv) Complicated Shapes Produced:

The complicated shapes that are not feasible by drawing process are easily produced by spinning process.

(v) No Investment is Die Making:

The spinning operation can be tried quickly and without any investment on new models as it requires no special tooling and die making.

(vi) No Finishing Operation Required:

The spun parts require no finishing operation like bending trimming etc.

Limitations of Spinning:

(i) More Time Required:

The time required to produce a cup is more in spinning than drawing process.

(ii) More Skill Required:

The spinning process requires more skill of operator as tool is manually pressed against the metal blank.

(iii) Not Suitable for Large-Scale Production:

The spinning process is not suitable for large-scale production, as it consumes more time.

Applications and Uses of Spinning:

1. Most of metals that are available in sheets can be spun. The most commonly used are: Aluminum, Copper, brass, stainless steel, carbon steel etc.
2. Molybdenum, titanium, and magnesium alloys are usually spun by hot spinning process.
3. Aluminum alloy up to 6 mm thick, low carbon steels and brasses up to 4.5 mm thick and stainless steel up to 3mm thick can be easily hand spinning.
4. The spinning process is used for producing cup shaped articles which are axis-symmetrical.
5. The spinning process is used to manufacture a small batch of articles where tools costs for a drawing process are uneconomical.
6. Spinning is particularly suitable for making deep conical parts.
7. Many complex articles and re-entrant shapes can be produced only by spinning.
8. Spinning can also be used to obtain hollow products with a diameter at the neck smaller than at the middle.

9. TYPES OF PRESSES

Presses are machines which apply pressure to a workpiece to change its shape and form. Used for a multitude of fabrication processes—e.g., die cutting, forging, and stamping—these machines can be used for processing a variety of materials (generally metal) across a wide range of industries.

There are numerous characteristics by which presses can be designed, constructed, and categorized. These characteristics include function (i.e., the fabrication process, such as die cutting presses), structure and design (e.g., screw press or gap press), and integrated controls (conventional vs. servo presses). Additionally, presses are available with, and categorized by, different types of actuating mechanisms—which produce and exert the force necessary to execute the fabrication process—including:

- Mechanical
- Hydraulic
- Pneumatic

Mechanical Presses

Mechanical presses employ different types of mechanical components to exert force, including hand cranks, knuckle joints, ball screws, lead or power screws, and rack and pinion drives. These presses can be either manually-operated—i.e., hand-powered by an operator—or electromechanically-operated—i.e., power is provided by a motor. Regardless of the power source, in all mechanical presses, the radial motion of the mechanical components is translated into linear motion which is, in turn, exhibited as a pressing force on the workpiece.

Of the types of presses available, mechanical presses have the simplest set-up and operation requirements, as well as offer high levels of control, precision, and accuracy as the mechanical components are set to a fixed stroke speed, length, and duration. Altogether, these characteristics enable mechanical presses to produce relatively cheap, fast, repeatable, and scalable results, which is essential in large-scale part production. Additionally, they pose a smaller environment risk compared to other types of presses as they do not require fluids for use.

Despite these advantages, mechanical presses are limited in some capacities, including by their fixed stroke lengths and the variable force produced throughout the duration of the stroke. These characteristics make them unsuitable for applications which may require variable stroke lengths—e.g., in open die forging—or the consistent application for force—e.g., in extrusion.

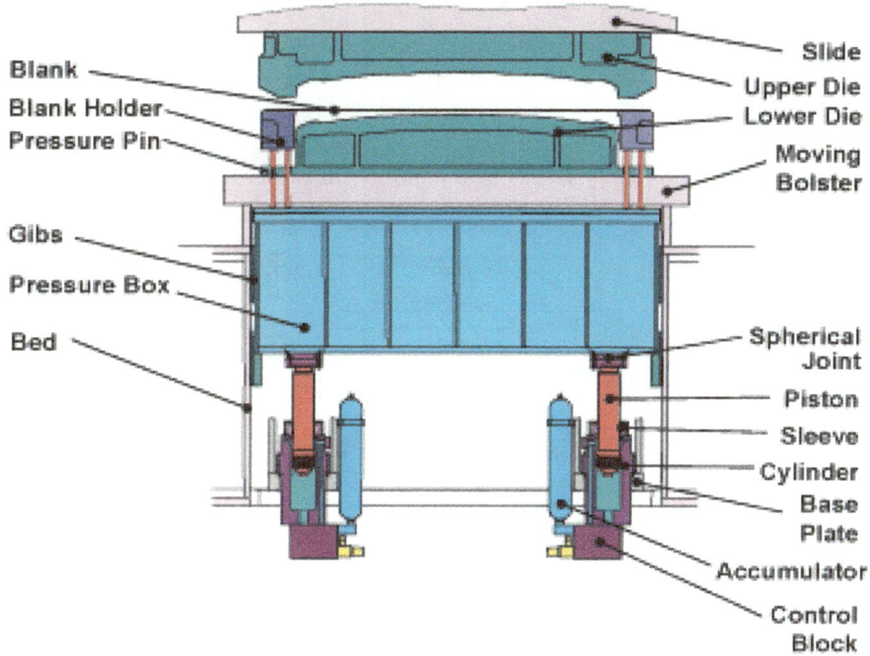

Fig.28 Mechanical presses

Hydraulic Presses

Hydraulic presses employ a hydraulic mechanism consisting of a hollow cylinder, piston, and power unit. The power unit moves the piston within the cylinder, which increases the internal fluid pressure and causes the cylinder to move along the axis of the piston. This motion creates a linear force which then extends or contracts the press's components, increasing or decreasing the amount of pressure applied to the workpiece.

While hydraulic presses are available in a wide range of sizes and capacities, the maximum force attainable by a particular press is pre-determined by its cylinder bore size and the operating pressures. Its ram speed is also restricted by the limitations of the power unit. Additionally, given the use of hydraulic fluid in the press's mechanism, there is an increased risk of environmental leakage.

Despite the above disadvantages, hydraulic presses offer several advantages over other types of presses, including greater force capabilities and consistent press force throughout the stroke range.

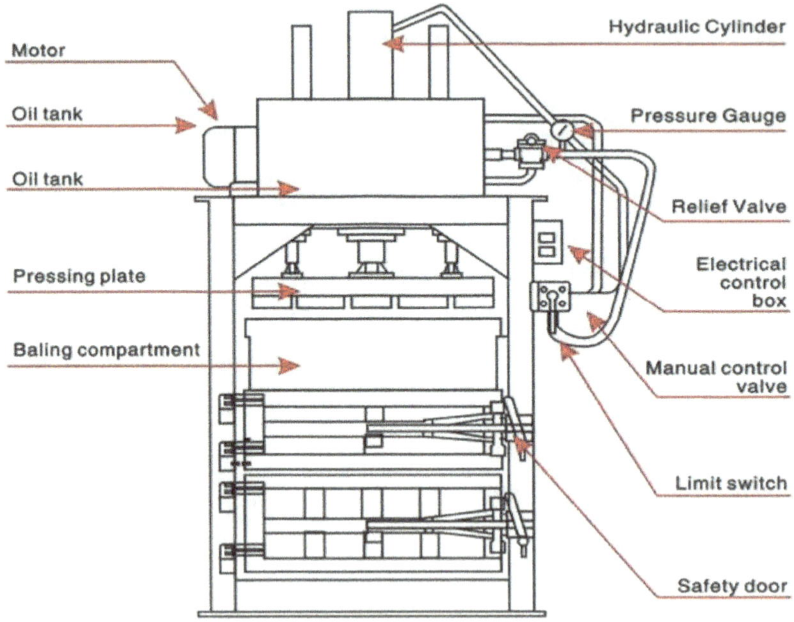

Fig.29 Hydraulic presses

Pneumatic Presses

Pneumatic presses employ a mechanism similar to that of hydraulic presses; except instead of compressing a hydraulic fluid, the pneumatic press mechanism works by compressing air to increase the internal pressure within the cylinder.

The advantages of employing pneumatic presses are the consistency of the press's force throughout the stroke range, higher working speeds, and suitability for extreme temperatures and explosion-prone applications. Despite these advantages, pneumatic presses, are also limited by their actuating mechanism. Like hydraulic presses, they are restricted both by cylinder size and compressor. Additionally, when using a pneumatic press, operating pressure must be maintained throughout the entire process to ensure proper operation and prevent operator injury.

Types of press tools

Press tools are commonly used in hydraulic, pneumatic, and mechanical presses to produce the sheet metal components in large volumes. Generally press tools are categorized by the types of operation performed using the tool, such as blanking, piercing, bending,

Blanking tool

1. Blanking is a punching operation in which entire periphery is cut out and the cut out portion required is known as STAMP OR BLANK.

- When a component is produced with one single punch and die where the entire outer profile is cut in a single stroke the tool is called a blanking tool.
- Blanking is the operation of cutting flat shapes from sheet metal.
- The outer area of metal remaining after a blanking operation is generally discarded as waste.
- Size of blank or product is the size of the die & clearance is given on punch.
- It is a metal cutting operation.

In blanking metal obtained after cutting is not a scrap or it is usable.

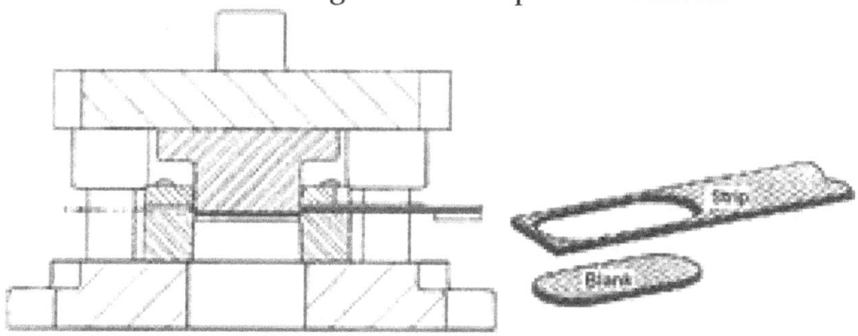

Fig.30 Blanking

Piercing tool

Piercing involves cutting of clean holes with a resulting scrap slug. The operation is called die cutting and can also produce flat components where the die, the shaped tool, is pressed into a sheet material employing a shearing action to cut holes. This method can be used to cut parts of different sizes and shapes in sheet metal, leather and many other materials.

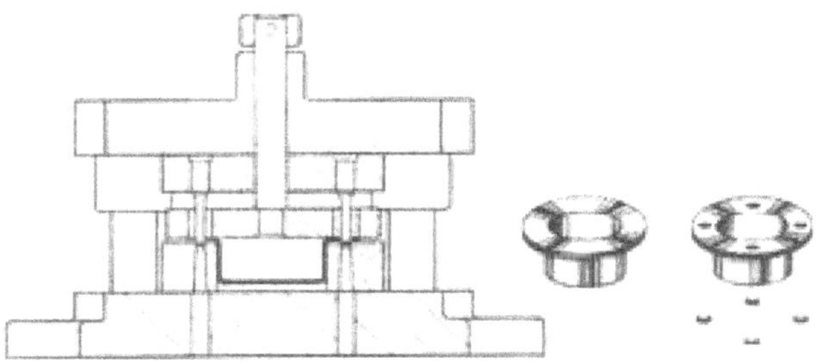

Fig.31 Piercing

Cut off tool

It is a shearing operation in which blanks are separated from a sheet metal .strip by cutting the opposite sides of the part in sequence.

Parting off tool

The difference between a cutoff and a parting is that a cutoff can be nestled perfectly on the sheet metal, due to its geometry. With cutoffs, the cutting of sheet metal can be done over one path at a time and there is practically no waste of material. With partings, the shape can not be nestled precisely. Parting involves cutting the sheet metal along two paths simultaneously. Partings waste a certain amount of material, that can be significant.

Trimming tool

When cups and shells are drawn from flat sheet metal the edge is left wavy and irregular, due to uneven flow of metal. Shown is flanged shell, as well as the trimmed ring removed from around the edge. While a small amount of Material is removed from the side of a component in trimming tool.

Shaving tool

Shaving removes a small amount of material around the edges of a previously blanked stampings or piercing. A straight, smooth edge is provided and therefore shaving is frequently performed on instrument parts, watch and clock parts and the like. Shaving is accomplished in shaving tools especially designed for the purpose.

Forming tool

Forming is the operation of deforming a part in curved profile. Forming tools apply more complex forms to work pieces. The line of bend is curved instead of straight and the metal is subjected to plastic flow or deformation.

Drawing tool

Drawing tools transform flat sheets of metal into cups, shells or other drawn shapes by subjecting the material to severe plastic deformation. Shown in fig is a rather deep shell that has been drawn from a flat sheet. It is an axial elongation through the application of axial force.

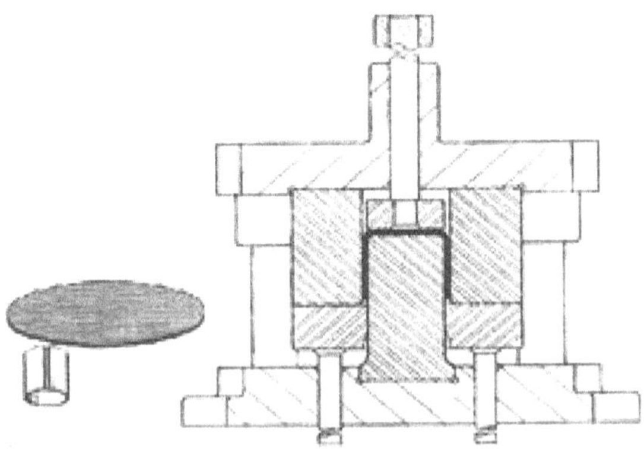

Fig.32 Drawing

Progressive tool

A progressive tool differs from a stage tool in the following respect: in a progressive tool the final component is obtained by progressing the sheet metal or strip in more than one stage. At each stage the tool will progressively shape the component towards its final shape, with the final stage normally being cutting-off.

Compound tool

The compound tool differs from progressive and stage tools by the arrangement of the punch and die. It is an inverted tool where blanking and piercing takes place in a single stage and also the blanking punch will act as the piercing die.

Combination tool In a combination tool two or more operations such as bending and trimming will be performed simultaneously. Two or more operations such as forming, drawing, extruding, embossing may be combined on the component with various cutting operations like blanking, piercing, broaching and cut off takes place- it can perform a cutting and non cutting operations in a single tool.

Cutting force in press tool

in general cutting force (CF) can be calculated using formula: CF =L x S x ζmax

cutting force will be in Newton(N) Where, L = Cut length in mm,(perimeter of profile to be cut) Ex: 40 mm square to be cut will have cut length of 160 mm

S = Sheet metal thickness in mm,

ζmax = Maximum shear strength of sheet metal MPa

Stripping force

Stripping force will be usually 10 to 20 % of cutting force

Press force

Press force will be cutting force with stripping force Press Force = Cutting force + Stripping force

UNIT – V

METAL FORMING PROCESSES - II

1. METAL EXTRUSION

Metal Extrusion is a metal forming manufacturing process in which a cylindrical billet inside a closed cavity is forced to flow through a die of a desired cross section. These fixed cross sectional profile extruded parts are called "Extrudates" and pushed out using either a mechanical or hydraulic press.

Most commonly extruded materials are Aluminium, Copper, Steel, Magnesium, and Lead.

Characteristics of Extrusion

- Able to create complex cross sections and will be uniform over the entire length of the extrudates
- Factors that affect the quality of extrusion are die design, extrusion ratio, billet temperature, lubrication, and extrusion speed
- Similar to any other metal forming processes, it can be performed either hot or cold, although the process generally is carried out at elevated temperatures in order to reduce the extrusion force and improve the ductility of the material
- Low cost due to reduced raw material wastage and high production rate
- Brittle material can be deformed without tear as it only exerts compressive and shear forces in the stock part
- Parts that are formed have excellent surface finish which minimizes post processing machining
- Metal extrusion tends to produce a favorable elongated grain structure in the direction of the material .
- The minimum wall thickness of ~1mm (aluminium) to ~3mm (steel) could be achieved

Cold Extrusion

The process of cold extrusion is carried out at room temperature or at marginally elevated temperature, with the assistance of Extruder and Extrusion Machines. These equipment / machines are specially developed on the basis of innovative Extrusion Technology. Cold extrusion can also be defined as the process of shaping of a cold metal by striking a slug. This striking is done with a punch (in closed cavity), which forces the metal in upward direction around the punch. This process is also called Cold Forging, Cold Pressing, Extrusion Pressing, and Impact Extrusion.

Cold extrusion is also defined as a compressive forming process (push-through), where the starting material is billet / slug and the process is carried out at the room temperature. During the cold extrusion process, deformation heating of the deforming material takes place at several hundred degrees. Deformation heating is the process of conversion of deformation work to heat. In general, a punch is used for applying pressure to the enclosed billet in the stationary die. This pressure can apply partially or completely, as per the requirements. On the basis of punch, die design along with the resulting material flow, we can classify cold extrusion into three major processes are Forward extrusion, Backward extrusion and Lateral extrusion.

Advantages of cold extrusion

- No oxidation process
- Higher strength due to cold working
- Good mechanical properties that are provided the temperatures created, are below the re-crystallization temperature
- Closer tolerances
- Good surface finish
- Fast extrusion speeds, in case the material is subjected to hot shortness Dimensional precision
- Net-shaped features
- Lower energy consumption
- Higher production rates
- Cleaner work environment.

Disadvantages

- High lubrication cost
- Higher loads
- Limited shape
- Complexity

Applications of Cold Extrusion

- The cold extrusion process is widely used to manufacture products from Copper, Lead, Tin, Aluminum Alloys, Titanium, Molybdenum, Vanadium, Steel, Zirconium, and Niobium.
- Some of the products that are developed using this process include various parts such as Collapsible Tubes, Gear Blanks, Aluminum Cans, Cylinders, Fire Extinguisher Cases, Shock Absorber Cylinders, and Automotive Pistons.
- Cold Extrusion is also used in the Automobile Sector, in various applications such as:
    - Injection technology
    - Engine control

- Fuel supply
- Automatic transmissions
- Seat technology
- Safety systems (restraint systems)

2. TYPES OF EXTRUSION PROCESS

Metal extrusion can be sub divided and grouped into the following categories depending on the direction of extrusion flow, the medium used to apply force, working temperature, etc.

- Direct Extrusion
- Indirect Extrusion
- Hydrostatic Extrusion
- Lateral or Vertical Extrusion
- Hot extrusion
- Cold Extrusion
- Impact Extrusion

Direct Extrusion (Forward Extrusion)

Direct extrusion, sometimes called Forward Extrusion is the most common type of extrusion. The process as shown in figure 2 below, begins by loading a heated billet (only for hot extrusion, discussed later) into a press cavity container where a dummy block is placed behind it. Then the mechanical or hydraulic ram presses on the material to push it out through the die. Then, while still hot, the part is stretched to straighten. This process is beautifully animated by Core Materials.

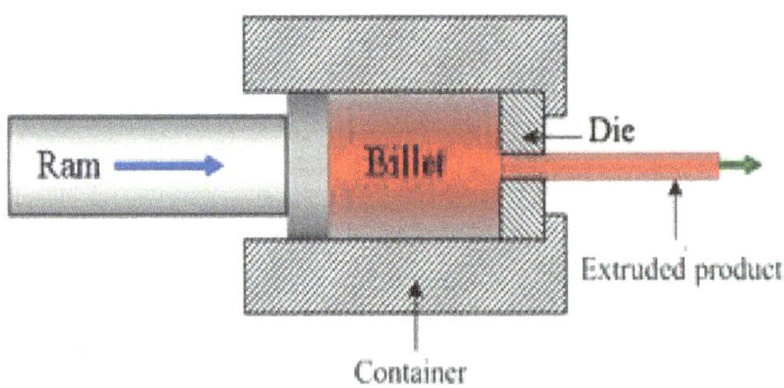

Fig.1 Direct extrusion

Under direct extrusion, the high friction caused by steels at higher temperatures is reduced using molten glass as a lubricant while oils with graphite powder are used for lubrication for low temperatures. The dummy block is used to protect the tip of the pressing stem (punch or ram) in hot extrusion. When the punch reaches the end of its stroke, a small portion of the billet called "butt end" cannot be pushed through the die opening.

Advantages of Direct metal extrusion

- No billet modification required
- Can be used for both hot and cold extrusion
- Simple tooling compared to other extrusion process

Disadvantages of Direct metal extrusion

- High force requirement due to friction
- Butt end left inside the cavity
- The force required to push the ram changes as the punch moves

Indirect Extrusion (Backward Extrusion)

In Indirect Extrusion, the die is located at the end of the hydraulic ram and moves towards the billet inside the cavity to push the material through the die. This is illustrated in figure 2 below.

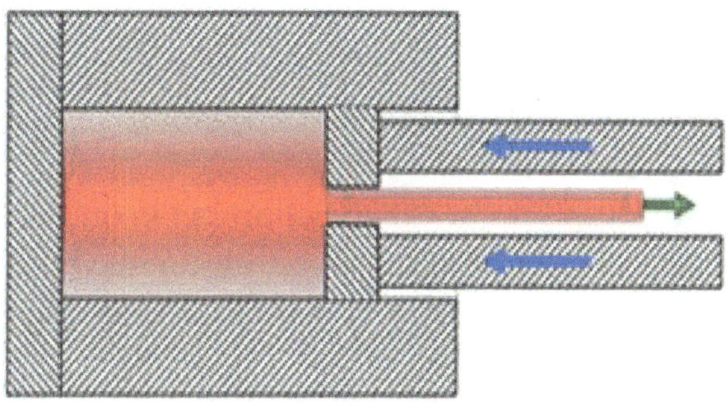

Fig.2 Indirect extrusion

This process consumes less power due to the static billet container causing less friction on the billet. However, supporting the extruded part is difficult when the extrudate exits the die.

Advantages of In-direct metal extrusion

- Less friction and less power used
- Can be used for both hot and cold extrusion
- Simple tooling compared to other extrusion processes

Disadvantages of In-direct metal extrusion

- Difficult to support the extruded part
- The hollow ram limits the load applied

Hydrostatic extrusion

In hydrostatic extrusion, the chamber/ cavity is made smaller than the billet and filled with hydraulic fluid which transfers the force from the ram to the billet as shown in figure 3. Although tri-axial forces are applied by the fluid, the pressure improves billet formability on the billet. Sealing the fluid must be considered at the early stages to avoid any leaking and reduced pressure issues.

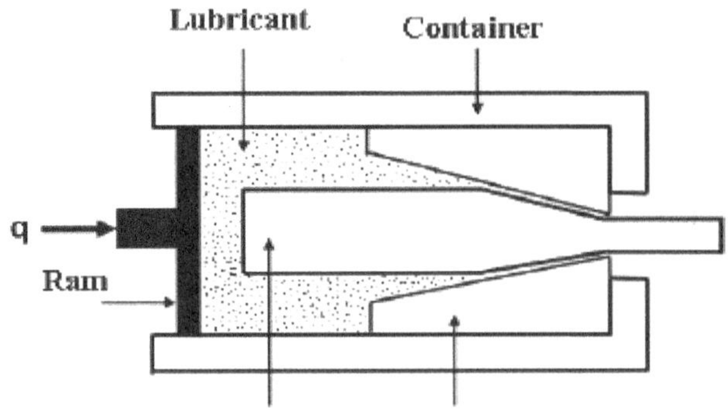

Fig.3 Hydrostatic extrusion

Although the hydraulic fluid eliminates the friction between the wall and the billet by isolating them, due to the specialized equipment requirement, high set up time and low production rate limits its usage in the industry in comparison to other extrusion processes.

Advantages of Hydrostatic metal extrusion

- Low power/force requirement due to no friction
- Fast production rates & high reduction ratios
- Lower billet temperature
- Even flow of material due to the balanced force distribution
- Large billets and large cross sections can be extruded
- No billet residue is left in the container

Disadvantages of Hydrostatic metal extrusion

- Billets needs preparing by tapering one end to match the die entry angle
- Only cold extrusion is possible
- Difficult to contain the high pressure fluid

Lateral Extrusion

In Lateral Extrusion the container is in vertical position as shown in the image and the die is located on the side. This process is suitable for low melting point material.

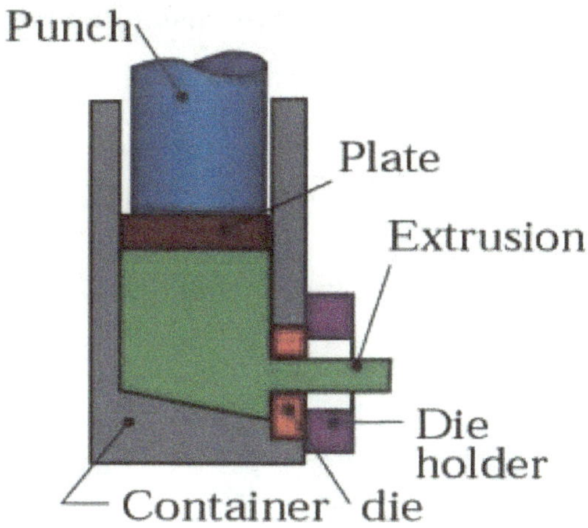

Fig.4 Lateral Extrusion

Impact Extrusion

Impact extrusion is part of cold extrusion category very similar to In-direct extrusion and limited to softer metal such as Lead, Aluminium and copper. As the schematic illustrates, the punch pushed down at high speed and extreme force on the slug to extrudes backwards. Thickness of the Extrude is a function of the clearance between the punch and the die cavity. The extrudatesare slide off the punch by the use of stripper plate.

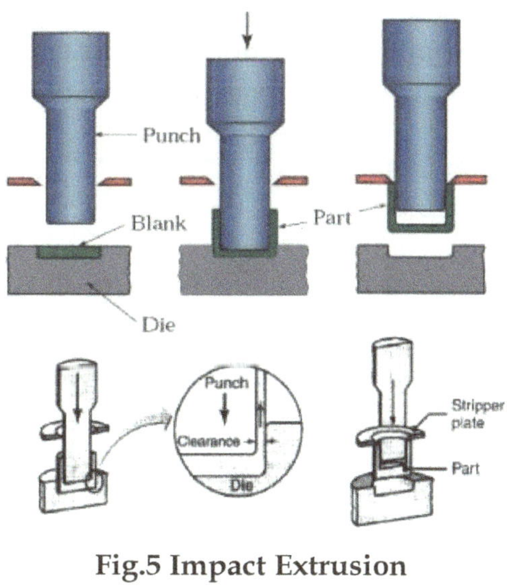

Fig.5 Impact Extrusion

For impact extrusions a mechanical press is often used and the part is formed at a high speed and over a relatively short strokeSince the forces acting on the punch and die are extremely high, tooling must have sufficient impact resistance, fatigue resistance and strength, for extruding metal by impact. Impact extrusion can be divided into the following three types by the flow of the material.

- Forward
- Reverse
- Combination

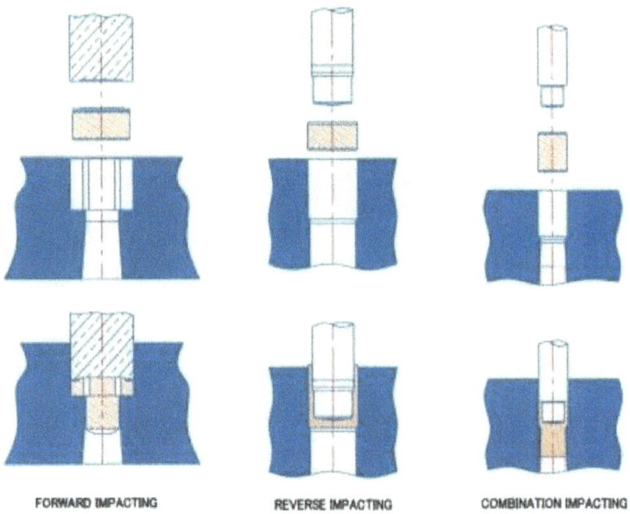

Fig.6 Types of Impact Extrusion

In forward impact extrusion, metal flows in the same direction that the force is delivered while it flows in the opposite direction in reverse impact extrusion. As the image shows above, in combination, the metal flows in both directions.

Advantages of impact metal extrusion

- Raw material savings of up to 90%
- Reduced machining times up to 75%
- Elimination of secondary machining operations
- Reduction in multi-part assemblies
- Improved mechanical properties for material strength and machining due to cold working of the material
- Significantly reduced total part costs up to 50%
- Hollow thin walled tubes, closed on one end, are often produced in manufacturing industry by backward impact extrusion.

Disadvantages of impact metal extrusion

- Produced as long as the part is symmetrical over the axis by which it is formed
- Many of the parts formed by impacting, in industry, will require further manufacturing processes, such as metal forging, ironing or machining, before completion

Hot Extrusion

Hot extrusion is one of the most popular method to develop objects having a fixed cross-sectional profile. This extrusion process is done at increased temperature, which keeps the materials from work hardening along with making the procedure of pushing the material through the die simpler. In general, a large amount of hot extrusions are done using horizontal hydraulic presses that belong to the range of 250 - 12,000 tons. Here, the pressure range is 30 to 700 MPa (4,400 to 102,000 psi), hence lubrication is needed. This lubrication can be done by glass powder in case of higher temperature extrusions and graphite or oil for all the lower temperature extrusions. Though the hot extruded materials are highly preferred, hot extrusion has couple of disadvantages associated with it that are the cost of extrusion machines and their upkeep.

Applications of Hot Extrusion

- The hot extrusion process is widely used to manufacture products from Aluminium, Copper along with their alloys.
- Some of the products that are developed using this process are Electrical Wires, Bars and Tubes.

3. EXTRUDING EQUIPMENT

Extrusion is one of the most basic and most useful processes in the bakery and confectionery industries. We've used it successfully with substances like bakery doughs, pasta, starches, crèmes, peanut butter, gums, and even pet food. Through the years, we've developed machines and mechanisms to accomplish a variety of extruding tasks. Some of these include:

- Complete extruders with forming heads and cutters to produce individual shaped pieces such as pretzels.
- Extruders to produce continuous ropes for various candy and pretzel products
- Custom forming heads and cutting devices designed to fit third party extruders

Combination Extruding Equipment

While the process of extrusion and the related equipment might seem basic, when it is used in the proper combination with additional co-extrusion and forming equipment, very unique products can be created. Some of the unique products we have experience with include:

- Co-extrusion equipment to combine materials such as filled candies & filled pretzels or candy.
- Unique forming heads for multi-strand twists, spirals, etc. for candies & baked snacks.
- Extruders to produce continuous, flat sheets or strips of dough, icing, or candy
- Tube extrusion

Tube extrusion

A tube is formed by pressing the extruded polymer through a circular die, cooling it in a water bath and then cutting it or winding it up. The twin screw extruder is used when the production of medical tubes requires a compound. An example is the incorporation of silver salts. These help to prevent sepses when placing a catheter or stent. Anticoagulants can be incorporated as well to reduce blood coagulation around the appliances.

The production of the tubes on a twin screw extruder enables the modification of the metering and the reduction of raw material stocks. The tubes can be extruded either directly from the twin screw extruder by means of a downstream melt pump, or the twin screw extruder can be used as a second extruder in a co-extrusion system. The material containing the active ingredient can be produced either as the outer layer or within the tube.

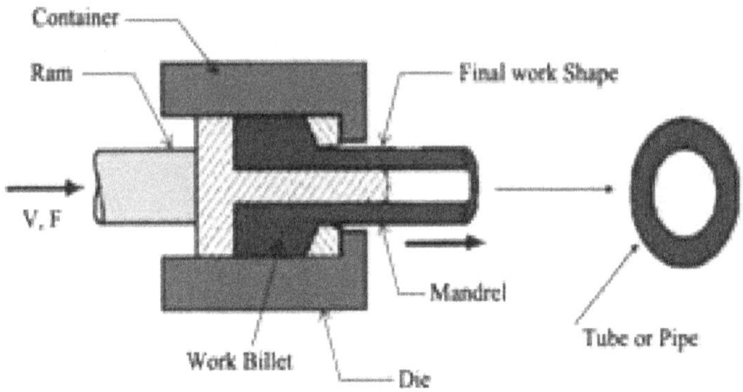

Fig.7 Tube Extrusion

Pipe Extrusion Process

Pipes are produced by horizontal extruding molten polymer through an annular opening in to several sizing, cooling devices that stabilize the final dimensions. The elements of the production line are:

1. Extruder
2. Die
3. Sizing device
4. Cooling bath
5. Puller
6. Cutter

Pipe Extrusion Die:

The role of an extrusion die is to form the molten material delivered by the screw in to a required cross section. The extrusion die can be classified as straight: through cross head and offset dies, base on their attachment to the extruder.

Sizing Equipment:

The extruded emerging from the pipe die usually shows variation of diameter, caused by variable of the rheological melt property. Thus the pipe has to be calibrated or sized while hot.

Cooling water through:

- During the sizing also the pipe gets cooled.
- Cooling is done by spraying water over the surface of the pipe.
- This length of the cooling through PE is longer than for PVC.

Pulling Machines:

- Pulling machines maintain uniform pipe size
- Pullers are either the caterpillar type of the gripping wheel type.
- The puller speed is controlled by electrical electronic or magnetic speed vartiators.

Controls:

- They provide automatic measurement and control of pipe size
- In one system the capacitance of the pipe is measure and is related to the thickness
- The sensing head moves around the pipe, showing when the pipe is out of round and when it is generally too heavy or too light.
- All this systems can be made of feedback to control the puller or the extruder drive.

Cutting Device:

- Circular say cutters with blade special designed for cutter plastics are used to cut the plastic in to specified length.
- Automatic cutting device are also used. They travel with the pipe to cut than after cutting go back to pick up and the next pipe length.
- Larger size are sawed to length

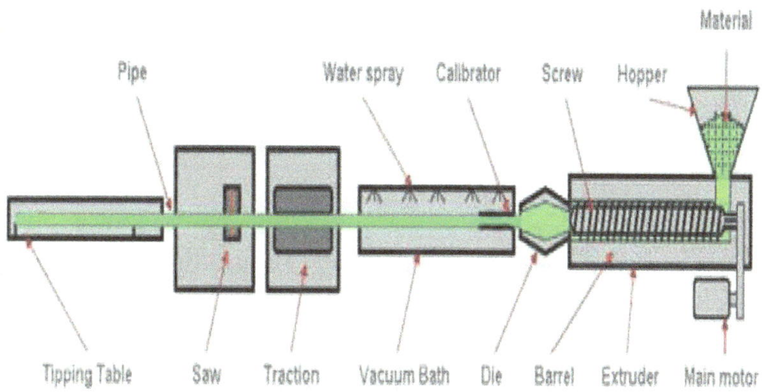

Fig.8 Pipe Extrusion

4. FORGING PROCESS

Forging is a manufacturing process involving the shaping of metal using localized compressive forces. The blows are delivered with a hammer (often a power hammer) or a die.

Operation:

Mainly forging consist two operations according to the force applied.

1) Drawing Down:

Drawing down is a process of elongate the length and reduce the cross section area of work piece. Simply in this operation, the length of work piece increases and the cross section area decreases. In this process, a compressive force is applied at perpendicular direction of its length axis. If a tensile force is applied to change its length at parallel to its length axis, this process is known as wire drawing.

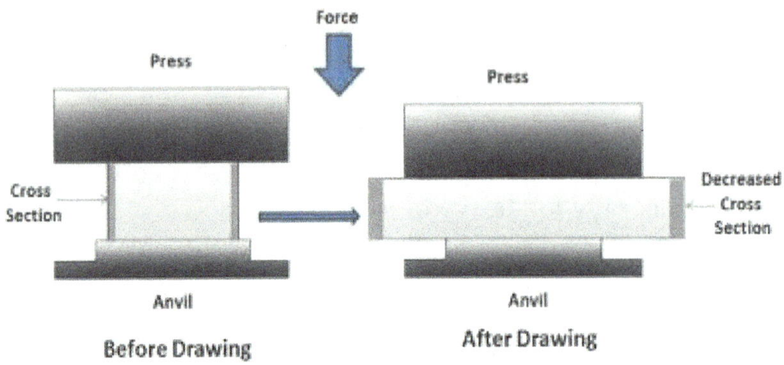

Fig.9 Drawing down

2) Upsetting:

Upsetting is just opposite operation to drawing down. In this operation, the length of work piece decreases and its cross section area increases. In this process, a compressive force is applied at parallel direction to its length axis.

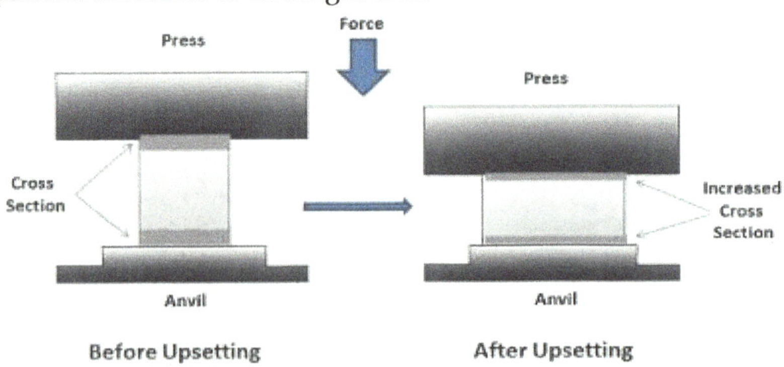

Fig.10 Upsetting

Types of Forging:

Forging may be classified into following ways.

1.) According to the Temperature:

a.) Hot Forging:

If the work piece is heated above its recrystallisation temperature and then force is applied, the process is known as hot forging. The recrystallisation temperature of most of metals is half of its melting temperature.

b.) Warm Forging:

Warm forging done at the temperature about 40 percent of its melting point temperature.

c.) Cold Forging:

Cold forging is done at room temperature. This process gives highest mechanical properties among all forging processes.

2.) According to arrangement of Die:

a.) Open Die Forging:

This forging process uses flat dies or no die. This is mostly hot forging process in which the work piece heated and hammered to obtain desire shape. This process produce rough product.

b.) Close Die Forging:

Close die forging is a process in which a set of impression die is used to produce desire shape on work piece. Mostly these products do not need any finishing process. This process required higher force.

3.) According to the Forging equipment:

a.) Smith forging:

Smith forging is traditional method of metal forming. It is open die forging method in which the work piece is placed on a stationary anvil and a hammer strikes and deform the work piece. The force is applied either by manually or power hammers.

b.) Drop Forging:

Drop forging is done in closed impression dies. An automatic hammer applied the force in series of blow by dropping action. In this type, a sudden applied force is used.

c.) Press Forging:

Press forging is similar to drop forging except it uses continuous squeezing type applied force by the hydraulic press. These processes increase the length and decrease its cross section of work piece.

d.) Machine Forging:

Machining forging is an upsetting process in which the cross section of work piece increases and length decreases.

Advantages and Disadvantages:

Advantages:

- Forging gives comparatively tougher product compare to **casting**.
- The fatigue strength and creep resistance of forge product is higher.
- Forge product has higher mechanical properties.
- Low cost operation.
- This process does not required special skill operator.
- Variety of shapes can be formed by this process.

Disadvantages:

- Higher initial cost for big forging presses.
- Secondary finishing process required in hot forging.
- It cannot produce complex shapes.
- Size is limited due to size of press.
- Brittle metal cannot be forged.

Application of Forging:

- Mainly forged products are used in every mechanical industry.
- **Turbine** rotor, generator rotor etc. are forged product.
- It gives higher fatigue strength so most of moving parts like crankshaft, camshaft **gears** etc. are made by forging operation.
- Cold forging is used to produce chisel, bolts etc.
- These are mostly used in hand tools and hardware manufacturing.
- It is used in ship building in various structure works.

5. FORGING METHODS

There are several forging methods with different capabilities and benefits. The more commonly used forging methods include the drop forging methods, as well as roll forging.

Drop forging

Drop forging derives its name from the process of dropping a hammer onto the metal to mold it into the shape of the die. The die refers to the surfaces that come into contact with the metal. There are two types of drop forging—open-die and closed-die forging. Dies are typically flat in shape with some having distinctively shaped surfaces for specialized operations.

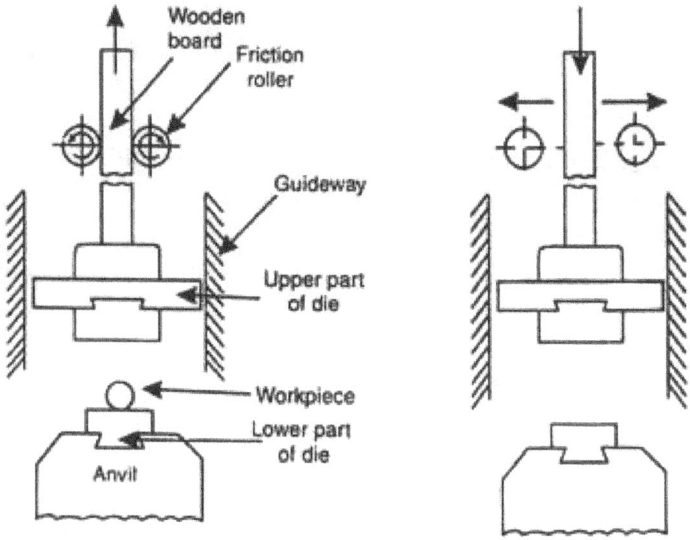

Fig.11 Drop forging

Open-die forging (smith forging)

Open-die forging is also known as smith forging. A hammer strikes and deforms a metal on a stationary anvil. In this type of forging, the metal is never completely confined in the dies allowing it to flow except for the areas where it is in contact with the dies. It is the operator's responsibility to orient and position the metal to achieve the desired final shape. Flat dies are used, with some having specially shaped surfaces for specialized operations. Open-die forging is suitable for simple and large parts, as well as customized metal components.

Advantages of open-die forging:

- Better fatigue resistance and strength
- Reduces chance of error and/or holes
- Improves microstructure
- Continuous grain flow
- Finer grain size

Closed-die forging (impression-die)

Closed-die forging is also known as impression-die forging. The metal is placed in a die and attached to an anvil. The hammer is dropped onto the metal, causing it to flow and fill the die cavities. The hammer is timed to come into contact with the metal in quick succession on a scale of milliseconds. Excess metal is pushed out from the die cavities, resulting in flash. The flash cools faster than the rest of the material, making it stronger than the metal in the die. After forging, the flash is removed.

In order for the metal to reach the final stage, it is moved through a series of cavities in a die:

1. **Edging impression (also known as fullering or bending)**
   The first impression used to mold the metal into a rough shape.
2. **Blocking cavities**
   The metal is worked into a shape that more closely resembles the final product. The metal is shaped with generous bends and fillets.
3. **Final impression cavity**
   Final stage of finishing and detailing the metal into the desired shape.

Advantages of closed-die forging:

- Produces parts up to 25 tons
- Produces near net shapes that require only a small amount of finishing
- Economic for heavy production

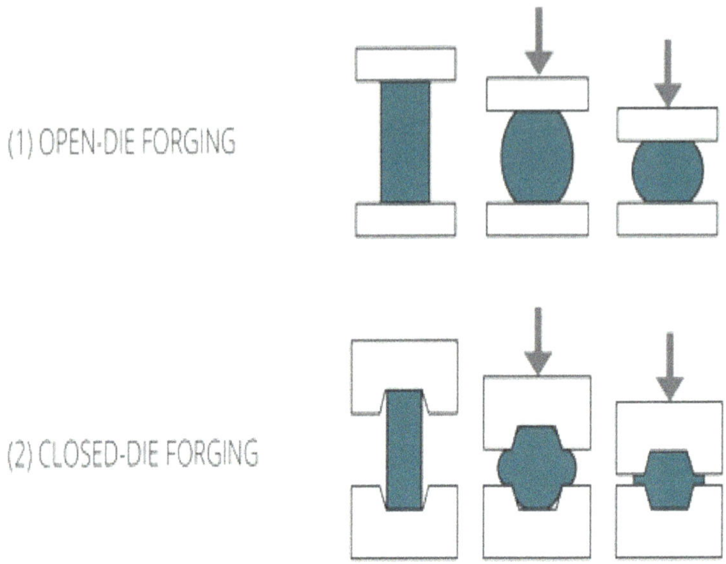

Fig.12 Open and closed die forging

Roll forging

Roll forging consists of two cylindrical or semi-cylindrical horizontal rolls that deform a round or flat bar stock. This works to reduce its thickness and increase its length. This heated bar is inserted and passed between the two rolls—each containing one or more shaped grooves—and is progressively shaped as it is rolled through the machine. This process continues until the desired shape and size is achieved.

Advantages of automatic roll forging:

- Produces little to no material waste
- Creates a favorable grain structure in the metal
- Reduces the cross-sectional area of the metal
- Produces taper ends

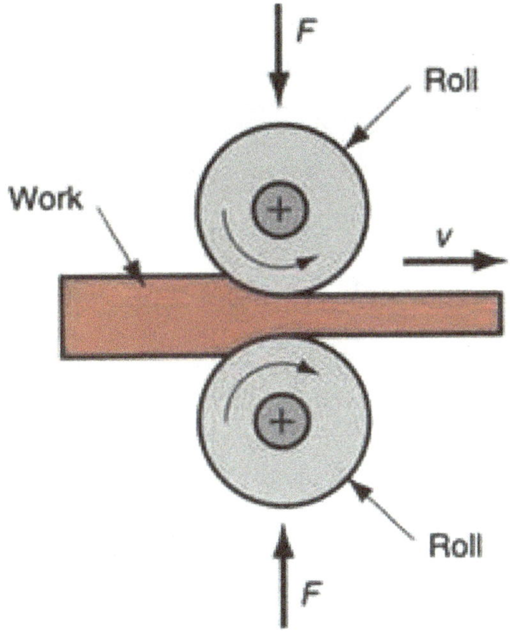

Fig.13 Roll and closed die forging

Press forging

Press forging uses a slow, continuous pressure or force, instead of the impact used in drop-hammer forging. The slower ram travel means that the deformation reaches deeper, so that the entire volume of the metal is uniformly affected. Contrastingly, in drop-hammer forging, the deformation is often only at the surface level while the metal's interior stays somewhat undeformed. By controlling the compression rate in press forging, the internal strain can also be controlled.

Advantages of press forging:

- Economic for heavy production
- Greater accuracy in tolerances within 0.01–0.02 inch
- Dies have less draft allowing for better dimensional accuracy
- Speed, pressure, and travel of the die are automatically controlled
- Process automation is possible
- Capacity of presses range from 500–9000 tons

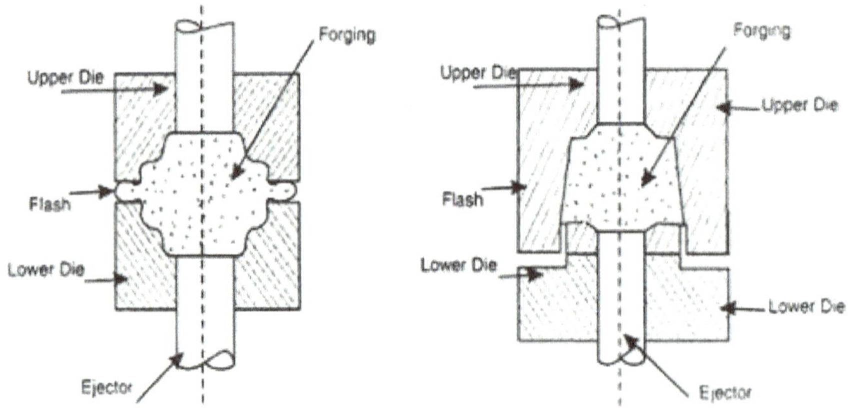

Fig.14 Press forging

Upset forging

Upset forging is a manufacturing process that increases the diameter of the metal by compressing its length. Crank presses, a special high-speed machine, are used in upset forging processes. Crank presses are typically set on a horizontal plane to improve efficiency and the quick exchange of metal from one station to the next. Vertical crank presses or a hydraulic press are also options.

Advantages of upset forging:

- High production rate of up to 4500 parts per hour
- Full automation is possible
- Elimination of the forging draft and flash
- Produces little to no waste

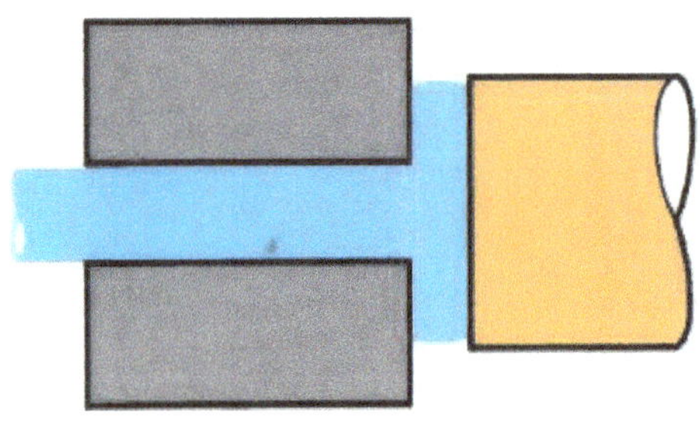

Fig.15 Upset forging

Automatic hot forging

In automatic hot forging, mill-length steel bars are inserted into one end of the forging machine at room temperature, and hot forged products emerge from the other end. The bar is heated with high-power induction coils to a temperature ranging from 2190–2370°F in under 60 seconds. The bar is descaled with rollers and shared into blanks. At this point, the metal is transferred through several forming stages that can be coupled with high-speed cold-forming operations. Typically, the cold-forming operation is left for the finishing stage. By doing so, the benefits of cold-working can be reaped while also maintaining the high speed of automatic hot forging.

Advantages of automatic hot forging:

- High output rate
- Acceptance of low-cost materials
- Minimal labor required to operate machinery
- Produces little to no material waste (material savings between 20-30% over conventional forging)

Precision forging (net-shape or near-net-shape forging)

Precision forging requires little to no final machining. It is a forging method developed to minimize the cost and waste associated to post-forging operations. Cost savings are achieved from the reduction of material and energy, as well as the reduction of machining.

Isothermal forging

Isothermal forging is a forging process where the metal and die are heated to the same temperature. Adiabatic heating is used—there is no net transfer of mass or thermal exchange between the system and the external environment. The changes are all due to internal changes resulting in highly controlled strain rates. Due to the lower heat loss, smaller machines may be used for this forging process.

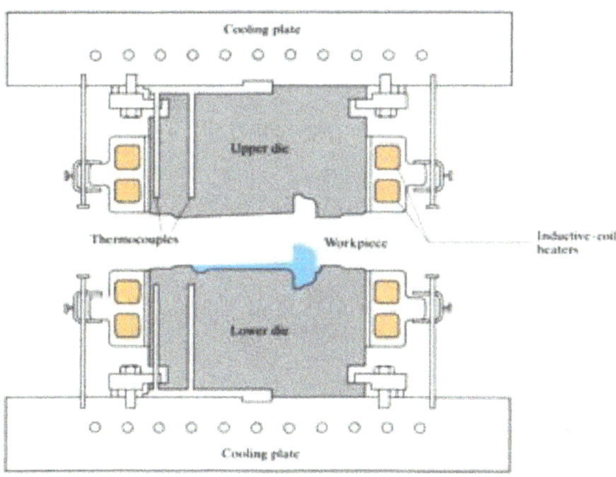

Fig.16 Isothermal forging

Forging Hammers

Forging hammers are used in the drop forging to form the metal between two dies. The first half of the die is attached to the anvil and the second part to the hammer. The material is placed in the lower die and then hammered with the upper one until the hot metal flows in all directions, filling the die cavity.

6. ROTARY FORGING

Rotary forging is an advanced and innovative metal forming technology that is used to manufacture a wide variety of mechanical components such as disks, rings, flanges and gears. During the cold rotary forging process the conical upper die continuously oscillates around the vertical machine axis at a constant revolution speed n. Simultaneously, the lower die pushes the work piece vertically at a constant feed rate v so as to cause it to be subjected to axial compression. Thus, after the work piece is pressed repeatedly for several times, the plastic deformation will be completed perfectly.

The principle of cold rotary forging is illustrated schematically in Figure. Different from conventional forging, the upper die in cold rotary forging is a conical die and can continuously oscillate around the vertical machine axis. Simultaneously the lower die pushes the work piece continuously so as to cause the work piece to be subjected to axial compression. Thus, when the work piece is pressed repeatedly for several times, the plastic deformation of the work piece will be completed perfectly.

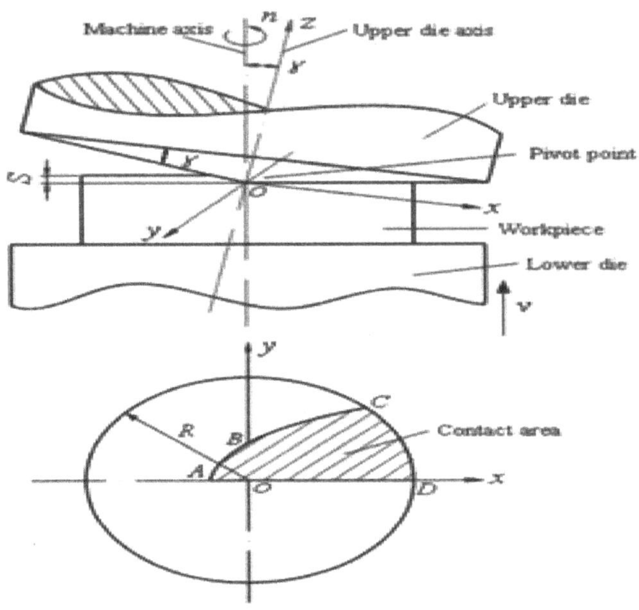

Fig.17 Rotary forging

Significant advantages have allowed cold rotary forging to have a variety of applications in many industrial fields such as automobile, machine tool, electrical equipment, cutting tool and hardware. Due to its potential advantages and wide application prospects, many studies have been done on the cold rotary forging process.

Rotary Forging Advantages as Compared with Conventional Forging

Technological:

- Greater deformations
- Force Reduction (5%-20%): less contact, less friction
- Greater Dimensional Accuracy: Near Net Shape
- Material Dimension Optimization: Burr Reduction/Elimination
- Greater surface finish quality (no cracks as a result of impact), incremental
- Material hardening and optimized grain structure
- Elimination of machining and welding operations
- Less noisy and polluting processes (without impact and sometimes without lubricants)

Economic:

- High flexibility —small modifications allow for new geometries
- Less investment —smaller presses
- Less expensive tooling —tooling geometry simplification
- Fewer tooling changes
- Fewer forming stages (2-4 stages in conventional forging)
- Fewer energy costs

7. METAL SWAGING

Metal swaging or radial forging involves the forging of a work piece by use of die that exert compressive forces by impacts that act around the work's circumference. In manufacturing industry, swaging is usually performed as a cold working process. In addition to the exterior geometry imparted by the impacting die, interior part geometry can also be forged by use of a mandrel.

In metal swaging manufacture, the work material is not completely restricted by the die. This is important to note when considering the metal flow that will occur during this forging process. Many impacts from each die may be required to completely forge the work to the geometry of the die and mandrel. In a typical metal swaging process, the work may receive several forging blows per second. These forging strokes are performed and coordinated by rollers that act in a cage. The cage rotates around the part, and the rollers, in turn, enact the motion of the die.

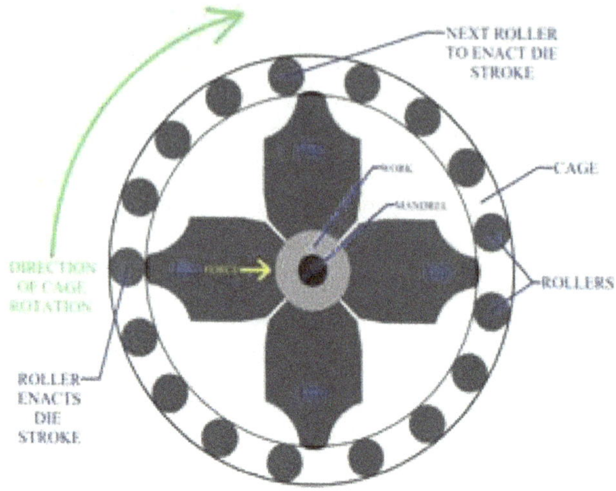

Fig.18 Metal swaging

Cold Forging

Cold forging is a manufacturing process where a bar stock is inserted into a die and squeezed with a second closed die. The deformation starts at room temperature and changes the shape and size of the initial part until it has assumed the shape of the die. At this final step, the part can reach a temperature of up to 250°C, since the friction strain rates are very high during deformation.

The process requires high-resistance dies and high-duty steel, usually reinforced by retaining rings. Even if these dies are quite expansive compared to hot forging dies, their life time is much higher, leading to a competitive cost per unit.

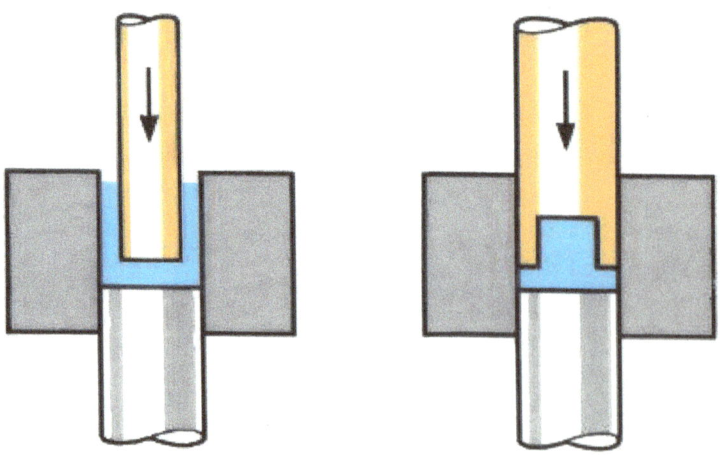

Fig.19 Cold forging

Hot forging

Hot forging is a metalworking process in which metals are plastically deformed above their recrystallization temperature, which allows the material to retain its deformed shape as it cools. Hot forging is accomplished using hydraulic or pneumatic hammers, presses, and other similar machinery that are used to compress the heated metal into its desired part shape. This is most often achieved using a custom die. The dies used in hot forging are custom-made to match the customer's part designs. Because hot metal is more pliable, this process allows for more intricate shapes than cold forging, making a vast array of part geometries possible. However, tolerances used in hot forging are generally not as tight as in cold forging.

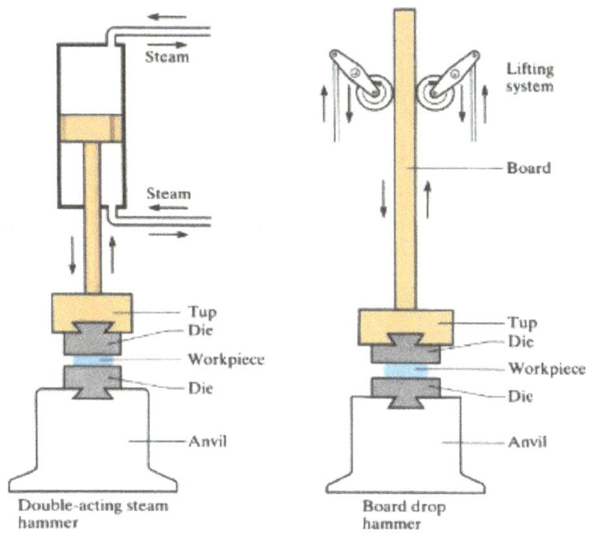

Fig.20 Hot forging